DANGEROUS YEARS

DAVID W. ORR

Dangerous Years

Climate Change,

the Long Emergency,

and the Way Forward

Yale UNIVERSITY PRESS

NEW HAVEN AND LONDON

Published with assistance from the foundation
established in memory of Amasa Stone Mather
of the Class of 1907, Yale College.

Yale University Press books may be purchased in
quantity for educational, business, or promotional
use. For information, please e-mail sales.press@yale.edu
(U.S. office) or sales@yaleup.co.uk (U.K. office).

Set in Janson Oldstyle and Futura Bold types by
Westchester Publishing Services.
Printed in the United States of America.

ISBN 978-0-300-22281-4 (hardcover : alk. paper)

Library of Congress Control Number: 2016939429

A catalogue record for this book is available from the
British Library.

This paper meets the requirements of
ANSI/NISO Z39.48–1992 (Permanence of Paper).

10 9 8 7 6 5 4 3 2 1

To David Crockett

CONTENTS

PREFACE

This book concerns the long-term political, economic, and social implications of climate destabilization. Time will tell whether the Paris Agreement reached in December 2015, COP-21, is the beginning of a serious global effort to cap off the worst that could happen. Even so, our climate and other Earth systems will not reach a new equilibrium for a very long time. By then the Earth may have become a different planet, what Bill McKibben calls "Eaarth," with a hotter and more capricious climate.

The numbers are daunting. In March 2016, the concentration of carbon dioxide (CO_2) in the atmosphere crossed the 402 parts per million (ppm) threshold, a 42 percent increase over the preindustrial level. The total of other heat-trapping gases measured in carbon-dioxide-equivalent units is perhaps another 50 to 70 ppm. As a result the temperature of the Earth is higher by one degree Celsius (or 1.7 degrees Fahrenheit), with perhaps another half a degree of warming in the pipeline due to the lag between what comes out of our tailpipes and smokestacks and the resulting climate-change-driven weather effects we experience. We will be fortunate to cap CO_2 levels at 450 ppm, hold the warming to two degrees Celsius, and tiptoe successfully around carbon-cycle feedbacks that could trigger catastrophic changes. But we have every reason of prudence, morality, and sheer survival to meet and exceed those targets as quickly as possible. We are learning that the climate system is complex and nonlinear—that is to say, unpredictable and wholly unforgiving of human error and tardiness of the day-late, dollar-short kind. We have initiated very large changes in Earth's atmosphere with a duration measured in centuries to millennia, but our institutions, organizations, systems of governance, economies, and thinking are geared to the short term, measured in years to a few decades.

On the other side of the equation is the rapidly growing technological capacity to power the U.S. and global economies by a combination of increased energy efficiency and renewable energy in various forms. Here there are good reasons for sober optimism, but the path ahead won't be easy. The physics of energy and the laws of thermodynamics are unmovable, as are the obscure facts of energy return on investment and power density. The energy scaffolding of the modern world was built on fossil fuels that are highly concentrated, portable, and relatively cheap. Renewable energy in its various forms is diffuse, more difficult to concentrate, more expensive, and has a lower power density and energy return on investment. Demographics and human behavior also compound the difficulties posed by physics. The world's population is now 7.4 billion heading toward a crest of perhaps eleven billion. Our material expectations and needs for mobility are higher than ever and still growing. And there is good reason to believe that we have already overshot the Earth's carrying capacity.

Against this backdrop, the odds of curbing the worst that could happen are said to be fifty-fifty, give or take. For perspective, no sane person would get in a car with those odds of a fatal accident.

In this book I consider some of the changes that must occur in order for the world to come through the dangerous years of climate chaos intact. My focus is not on technological transformation, but on the deeper changes of governance, economy, education, and heart that underlie our predicament as both the cause of our plight and the solution. Both hardware and software changes are necessary, but neither is sufficient in itself. And both must be recalibrated to a longer horizon. Our problems are compounded because climate change is just one of several interrelated threats to our common future. Each of these threats is global, permanent, and symptomatic of deeper flaws embedded in our systems of governance, politics, economy, sciences, demographics, and culture. Together they represent a system crisis that will span centuries. "We are faced not with separate crises," wrote Pope Francis in his encyclical *Laudato Si'*, "but rather with one complex crisis which is both social and environmental."

I do not believe that we are fated to destroy the Earth by fire, heat, or technology run amuck. But if there is a happier future it will come down to this: *to act with compassion and energy*, our hearts must be in it; *to act intelligently*, we must understand that we are but one part of an interrelated global system; *to act effectively and justly*, we must be governed by accountable, transparent, and robust democratic institutions; and *to act sustainably*, we must live and work within the limits of natural systems over the longterm.

In other words, we must learn to solve for a pattern that includes human contrivances of economy, governance, education, technology, society, culture, and behavior embedded in the ecosphere of air, land, waters, other species, and complex biogeochemical cycles. The rub is that we're not very good at solving systems problems that are large or may endure for an extended period. First we deny the problem; then we dawdle; and, when finally forced to act, we tend to overlook the underlying structural causes and fiddle with small changes at the margins that often have unforeseeable and counterintuitive effects.

For these and other reasons, the necessary changes will most likely begin in neighborhoods, cities, states, regions, and networks of global citizens. They must start at a manageable and comprehensible scale by a process of trial and error. And they must cascade up to change the larger systems of governance and economy. In time such small efforts will "change everything," as Naomi Klein puts it, including our economy, consumption habits, expectations, governance, distribution of wealth, and the practice of democracy. Computer scientist James Martin, in *The Meaning of the 21st Century*, expresses the belief that we need "another revolution, putting into place the desirable management, laws, controls, protocols, methodologies, and means of governance." Economist Herman Daly believes that the necessary changes will "require something like repentance and conversion." My own view is that both and more will be necessary to navigate the dangerous years ahead.

Finally, a sustainable, decent, equitable, and genuinely democratic society cannot long exist as an island in a global system ruled by threats, violence, and the prospect of nuclear war. Someday something will go horribly, horribly wrong. In the meantime the war system sucks everything dry, squanders people and treasure alike, desiccates the practice of democracy, corrupts our habits of speech, and clouds our awareness of better possibilities. Not least, the habit of violence predisposes us to think of nature as merely another thing to be conquered. My point is that there can be no sustainable and just economy and harmony between humans and natural systems in a society ruled by fear, threats, violence, and war. A house so divided will fall.

ACKNOWLEDGMENTS

This book owes a great deal to friends and colleagues who helped me over many years. To all, I am indebted. I am especially grateful to those who read and commented on earlier drafts or chapters, including Bill Becker, David Benzing, Ed Long, Elizabeth Kolbert, Bob Longsworth, Tom Lovejoy, Carl McDaniel, Daniel Orr, Julie Schor, Rumi Shammin, Gus Speth, Edward O. Wilson, and George Woodwell.

I thank Jean Thomson Black at Yale University Press for her encouragement and my copyeditor, Julie Carlson, who affably and ably transformed a manuscript into a book.

I also acknowledge friends and colleagues at Oberlin who continue to make this institution one of America's greatest colleges—particularly President Marvin Krislov and senior staff members Bill Barlow, Eric Estes, Mike Frandsen, Ben Jones, Jane Mathison, Kathryn Stuart, and Ron Watts. I also thank Steve Mayer and John Petersen who are great colleagues and good friends.

My colleagues in the Oberlin Project have also made this work easier by doing theirs so ably. Thanks go to Heather Adelman, Sean Hayes Cullen Naumoff, and Sharon Pearson.

Most important, I thank my wife, Elaine, for enduring, listening, understanding, and grounding the author and this book in music, community, and grandkids. To all I am indebted more than words can say, and the gift must move.

We ain't much, Lord, but we's the best we got.

—David Grimes, Copper River guide

Lonesome Dove, Texas, in Larry McMurtry's telling, is the most woebegone of all woebegone towns. It is flat, hot, and dusty, a dreary speck of a cow town in the wastelands of West Texas. A slick cowboy on the run (Jake Spoon) rides into town one day telling tales of a place that is green with beautiful rivers, forests, and mountains—a lush paradise called Montana. Mesmerized, the boys in the saloon decide to go to Montana, cows and all. How far is it? No one knows. Neither do they know anything of the rivers they'll have to ford, the vast lonely spaces and chasms ahead, the hungry bears and cougars that will see them as dinner, the mountains they will have to cross, or the well-armed and hostile tribes standing between them and their dreams. There is no AAA to chart the course and no highways, motels, restaurants, bars, Global Positioning System (GPS), or emergency roadside services. In a word, they are clueless.

Most epic journeys, whether in history or fiction, are rather similar. After a promising start crossing the Red Sea, Moses and the Israelites took a forty-year detour through the wilderness only to discover that the Promised Land was fully occupied by the Canaanites, who all things considered preferred to stay put. If they'd known the difficulties in advance, would the Israelites have fled the land of Egypt? There was certainly a case to be made for staying on the banks of the Nile, making their peace with the Pharaoh, and learning to worship the sun god instead of a more jealous and demanding Yahweh. If only Odysseus had foreseen not just the trials of the ten-year-long battle for Troy but also those of another decade to get back home to Penelope, he might have found Agamemnon's pique over

Paris's stealing Helen less aggravating. Or had the Joads known about the welcome committee and accommodations waiting for them in California they might have decided to wait out the Dust Bowl in Oklahoma, or head east to the Ozark or Appalachian mountains.

Some journeys, such as Don Quixote's quest in service to his lady Dulcinea, are a mixture of fantasy, high thinking, low comedy, tragedy, sorrow, laughter, and irony. Others, like Karl Marx's vision of communism, did not end as comedy, although that one certainly had its ironies and tragic elements. Marx imagined a proletarian paradise without capitalists or governments, in which workers would work in the morning and write poetry in the evening, all while flourishing in classless amity. In real life, however, the journey detoured through a wilderness of gulags, killing fields, and "great leaps" that would kill millions of capitalists, proletarians, and peasants alike and end badly.

Ever since the human diaspora out of the plains of Eastern Africa, we have been wanderers dreaming of someplace better. We travel sometimes as refugees like the Joads, or as opportunity seekers like the Israelites, or as dreamers like the cowboys in Lonesome Dove. Sometimes, like Chaucer's pilgrims on their way to Canterbury or Muslims going to Mecca, we go in search of enlightenment and salvation of one kind or another. And sometimes we go as crusaders to avenge or plunder, or just for the thrill of it. Other times, like Thoreau, we travel at home. We are a restless lot driven either by the lure of opportunity or desperation to get out of harm's way.

The talk around the saloon these days is increasingly of the latter kind of restlessness. In the two-centuries-long blowout called the Industrial Revolution, we burned through most of our cheap fossil fuels and easily accessible minerals, thereby jolting the climate out of its 12,000-year equilibrium that geologists call the Holocene. Having grown from a population of a billion now heading for eleven billion and having squandered half or more of our forests, soils, and our ocean resources, some have proposed that we migrate to a green, lush place called sustainability. It is said to be a paradise of solar collectors, wind machines, LED light bulbs, smart grids, recycling, organic agriculture, pure waters, green cities, circular economies, and end-less opportunities spawned by "green growth."

Many corporate CEOs, bankers, university presidents, heads of state, mayors, foundation presidents, and others proclaim their commitment to making their various enterprises sustainable, without saying what that means—mostly, one suspects, because they haven't thought much about it. It is possible, however, that they have thought about it at least enough to know that thinking further and deeper would raise troubling and possibly career-ending questions. Others, having surveyed the damage, are less sanguine about our prospects and propose that some of us blast off like dandelion fluff to another planet or to a station to be built in space, leaving the mess we've made behind to the less mobile.[1]

Whatever the purposes or the stated destination, the journey to sustainability will likely resemble other quests, pilgrimages, and transitions. It will be longer and more difficult than we've been told or can possibly foresee. Its leaders will be sinners and a few saints who will exhibit every known human trait of ambition, vanity, fear, lust, blindness, arrogance, as well as humility, foresight, wisdom, empathy, and nobility. In some distant time, the historical record will reveal a mixture of irony, paradox, comedy, tragedy, and fate. But with foresight, hard work, and luck, it could also be a story about the transition to an ecologically sustainable world. If only it was just as simple as deploying better and smarter technology like electric cars or photovoltaic panels, it would be an easier, shorter, and more certain journey. The facts, however, suggest otherwise.

We, and subsequent generations, will travel through uncharted territory with challenges at a scale, import, complexity, and duration that humans have never experienced. The collective 7.4 billion of us are starting out as a rag-tag and contentious lot. We do not necessarily agree about the destination or even whether there is such a place as Montana or, if there is, whether we can make it or not. We are beginning the march under different banners, singing different hymns and chanting different slogans. We do not much like each other, and we are armed to the teeth. A very, very few intend to fly to Montana in their private jets; a few others wish to go in a chauffeur-driven limousine. Many more are going by bus, but most are traveling on foot. Some intend to make lots of money on the journey, but others can hope only to survive. Without some method of rationing, food and water along the way will become scarce. The topography ahead will become hotter and more barren, and nature more capricious than we've ever known. There will be disagreements about which tools to bring and how best to use them. There will be no easy consensus on even how the procession will be governed or whether it can be governed at all. Further, there will be few reliable scouts and no signposts, accurate maps, or GPS to guide the way. So, from time to time we will become disoriented and lose our way. We will have to learn how to read unfamiliar signs, reorient ourselves, get our bearings, and sometimes backtrack to find our way again. No Las Vegas odds-maker would bet on us. But there is no agreement on whether our long, long odds this should be widely known or concealed by some unnamed, self-anointed elite.[2]

There is a great deal that we do not know, but we do know that the centuries ahead will demand the most extraordinary things of us. They will require leadership at all levels with an improved commitment to fact, data, and logic, that is, science. Of scientists and science, it will require a more consistent and stronger commitment to the kinds of knowledge that promote human betterment, ecological restoration, justice, and peace rather than weapons, manipulation, and frivolous consumption. It will require sacrifice and steadiness of people who must learn how to be good citizens and better neighbors. Around future campfires on the dark nights ahead they will need to remember their histories and recall stories that remind, inspire, energize, and console. They will need authentic hope, not wishful thinking, and worthy and practical visions, not ideologies. Above all, they will need steady, knowledgeable, and unflappable leaders with clear heads and compassionate hearts.[3]

All metaphors have their limitations and this one is no exception. There is no Montana to which we can escape. We will have to stay put and make the journey to sustainability where we are. Moreover, unlike the boys in Lonesome Dove, we don't have a choice because business as usual, by which I mean dependence on fossil fuels, economic inequalities, overconsumption, public manipulation, and militarization, will prove fatal to civilization, possibly sooner than later. Finally, the success of the journey to sustainability depends a great deal on how it begins, how it is organized, and whether people know in advance that this will be a long and difficult transition— not an afternoon outing—and pack accordingly. It is a long way from Lonesome Dove to Montana.[4]

one

The world is not a pleasant little nest made for our protection, but a vast and largely hostile environment, in which we can achieve great things only by defying the gods; and that this defiance inevitably brings its own punishment. It is a dangerous world, in which there is no security, save the somewhat negative one of humility and restrained ambitions. It is a world in which there is a condign punishment, not only for him who sins in conscious arrogance, but for him whose sole crime is ignorance of the gods and the world around him.

—Norbert Weiner

This is the first century in which humankind could terminate its own existence with technology that goes wrong.

—James Martin

A specter haunts our common future. It takes many forms but is of our own making—an offspring of our helter-skelter rush into the modern world. We have unwittingly condemned ourselves, and many generations to come, to live with the consequences of remarkable powers, recently acquired, that we did not and do not know how to tame. In the euphoria generated by rapid technological change, our growing mastery over nature, and expanding economies, we unleashed jinns, created monsters, tempted fate, and rushed in where angels would fear to go. Now we must master the art and science of containing the consequences of our actions, not for a few years or decades but over centuries. Nuclear wastes, for example, must be managed for tens of thousands of years. Even were we to instantly stop emission of carbon and other heat-trapping gases, the effects of what we've already done will still affect the climate thousands of years from now. Some chemical wastes will remain toxic for hundreds to thousands of years. The loss of species is forever. We have adversely changed the atmosphere, lands, forests, and waters of the planet. We know, or should know, these things and we must clearly understand their implications. But we know better than we act, and that gap between what we know and how we behave has given rise to the long emergency that lies ahead.

Over the five thousand years of human history our ancestors in some places did considerable damage to forests, soils, and biological diversity. But the damage was done in patches and in time the effects were repairable and/or occurred on a small enough scale that they did not diminish the slow but steady demographic and material expansion of humankind. Until quite recently an "empty" Earth had plenty of room for migration to less populated regions. The closing of the U.S. frontier in 1890 marked a symbolic transition to a very different and "full" world in which the last "empty" lands on the five continents were mapped, occupied, and exploited.

This outcome was foreordained. The creation of the Royal Academy of Sciences in London in 1660 and the publication of *The Wealth of Nations* in 1776 heralded the onset of powerful paradigm-changing ideas that would shift the trajectory and accelerate the velocity of history in irrevocable ways. By the middle of the twentieth century the marriage between science and economy had been consummated in the Manhattan Project, which ushered in the atomic age. Science and its prolific and powerful but blind offspring, technology, became the driver of modern economies.

The quickening pace of change now threatens to upend virtually everything humans have taken for granted, including our very humanity. Everything has changed but our manner of thinking, which remains tribal, insular, and myopic. At the scale of the individual, what is assumed to be normal masks stunning and unprecedented changes in the human condition. Readers over the age of sixty have lived through two and a half doublings of the human population and were the primary beneficiaries of the global bonfire that consumed

95 percent of all the oil ever burned (which is also only slightly less than all that will ever be burned). The fossil-fuel age changed more than the chemistry of the atmosphere and oceans. It drastically changed how we think about distance, time, work, and nature as well as what we think about. And in the briefest moment of history we came to believe that the miraculous and extraordinary are merely normal.

The present emergency has been a long time in the making. In *Critias*, Plato noticed that overgrazing had stripped the forests and soils from the hills of Greece, leaving the land as if it were the "skeleton of a body wasted by disease." In 1864 George Perkins Marsh wrote that "man is everywhere a disturbing agent. Wherever he plants his foot, the harmonies of nature are turned to discords." He wrote at a time when the world population was slightly more than one billion. A few decades later, in the United States and elsewhere, the pace of deforestation and soil loss triggered the Progressive conservation movement led by Gifford Pinchot and Theodore Roosevelt. Until the middle of the twentieth century, the discussion about "man in nature" mostly concerned growing human pressures on resources (soils, wildlife, and forests) and the pollution of air and water. But in the 1950s, as the world's population crossed the two billion mark, voices of alarm became more urgent, expressed in bestselling books by prominent conservation leaders including Fairfield Osborn and William Vogt.[1] Rachel Carson's 1962 book *Silent Spring* would also shift attention to qualitative changes that came with what she called the "Neanderthal age of science," in which the careless use of chemicals threatened the very foundations of life. Her book echoed public concerns about the science that had led to the creation of nuclear weapons. Similarly, good reasons exist for concerns about the effects on human health and reproduction of a century of promiscuous chemistry.[2]

In the early 1970s, issues of resource conservation, population growth, science, and careless technology were folded into a larger analytical framework. The advance of ecology, the development of systems dynamics, and the use of satellites and computers as research tools extended the capacity to research and model complex ecological, economic, and human interactions at a global scale over long periods. The behavior of complex systems is often counterintuitive. Complex systems are nonlinear, with emergent properties as well as long lead and lag times between cause and effect, and they can change abruptly from one state to another. None of this was evident in the mechanistic worldview of Newton or to the founders of the American Republic.[3]

The systems perspective also revealed two glaring and well-documented flaws in Western thinking. The first was "reductionism"—the tendency to reduce problems to their component parts and thereby isolate them from their larger context. Two hundred years ago reductionist science worked wonders while its dangers were concealed because the scale was still small and because the effects were not immediately apparent. The second flaw was a bias toward the short-term, which tends to redirect our attention from ends to means. Neither flaw appeared to be very important when the pace of change was slower, the stakes were lower, and the sheer mass of civilization was much smaller.

From a systems perspective, the challenge of sustaining civilization requires calibrating—and recalibrating, in perpetuity—the energy, water, and material consumption as well as the ecological effects of seven and soon, perhaps, eleven billion humans given the complex realities of ecology, thermodynamics, the biogeochemical cycles of the Earth, and ethics on a planetary scale. The transition goes beyond simply improving our technology, although deployment of better technology will be necessary. It will require developing and sustaining the extraordinary clarity, foresight, and wisdom needed to make corresponding changes in governance, politics, economy, and culture. None of this will be possible without a softening of the more violent and reptilian aspects of human psychology and behavior, which in turn will require a deeper understanding of our nature and evolution.[4]

Two hundred thousand years ago, a mere blink in geologic time, our ancestors walked out of the forests onto the plains of Africa as

Homo sapiens. The name implies not just consciousness but also an emerging capacity to think about thinking. Early in our evolutionary journey, hominids acquired hands with opposable thumbs, stereoscopic eyesight, and very large brains. Compared to other animals, they were slow, weak, and without formidable teeth and claws. But they were very smart and possessed gifts for language, for toolmaking, and for building culture. Migrating out of the cradle of Africa, they inhabited the farthest corners of the Earth. They formed tribes and societies, and eventually created cities and nations. In the process humans learned to farm, write, calculate, invent, build, philoso-phize, govern, write poetry, compose music, and create diverse cul-tures. So equipped, in time humans became the dominant geologic force on Earth in the age now called the Anthropocene. It hap-pened in fewer than five thousand generations, and the most conse-quential of the changes occurred in the past five. As humans spread over the Earth, we fractured ourselves by ethnicity, culture, his-tory, religion, wealth, nationality, and ideology. As a result we often failed to see our various journeys as parts of a single evolving enter-prise.[5]

Whatever our differences and similarities, one fact stands out: humans everywhere are, for the most part, fast thinkers but slow learners. We have difficulty understanding exponential growth and the interconnectedness of things. We flounder when contemplating the long term. From Aristarchus of Samos, circa 280 BCE, it took us until the seventeenth century to understand that the Earth is not the center of the universe. The Church would offer only a grudging apol-ogy to Galileo for its censorship and only 359 years later, in 1992. Darwin published *The Origin of Species* in 1859, but many people in the United States in particular still gag on the idea that we are but a twig on a branch attached to the tree of life, kin in varying degrees to all the life that ever was. It took a half century for scientists to accept Alfred Wegener's theory that what appears as solid ground isn't so solid after all, but instead continent-sized plates floating on a molten sea below. Neoclassical economists still ignore the inconvenient laws of thermodynamics, preferring to build their mathematical castles in thinner air where the gravitational laws of entropy presumably do not apply. For most members of Congress and many in the larger society, the science of ecology is a dark void utterly beyond comprehension. The idea that the atmosphere is a system governed by physical laws of chemistry, biology, and physics is still hotly disputed for vaporous reasons. And we still struggle to understand the teachings of the wisest ever to walk among us, clinging instead to our fears, ideologies, and tribal idols.

For all of our pretensions to the contrary, the record shows that we are a precocious and violence-prone offshoot of the hominid line that in the briefest moment of geological time has come to dominate the planet, overwhelm its biotic processes, destroy entire forests, carelessly intrude in the elementary structures of matter and life, slaughter fellow creatures, pollute the skies and waters, fill virtually every ecological niche, and destabilize the climate for thousands of years to come. Biologist Lynn Margulis once called us a rogue species. It is a charge that is difficult—perhaps impossible—to refute. But I write in the belief that we can do better; that we are capable of improvement and self-correction. I am convinced that we can discipline our unruly human appetites and learn to act with compassion, wisdom, and foresight—that we can mature to the fuller stature of our self-advertised status as *Homo sapiens.*

It is also written in the conviction that we will have to act fast, much faster than ever before, to head off catastrophe from three perils that we have brought on ourselves. The most ominous, even decades after the end of the Cold War, is the ongoing threat of nuclear war that once sparked could cascade into a global holocaust. Since the entire system operates in secrecy, we have no idea whether the existing national and international firebreaks could contain a regional nuclear war or not. Neither do we know whether the existing safety systems would be adequate to prevent an accidental launch of nuclear weapons or a nuclear cascade that begins with an act of terrorism. What we do know is that we've been very lucky so far and that the global effects of even limited nuclear war would be catastrophic. The danger of nuclear war did not disappear at the end of the Cold War in 1990. Ten thousand or more aging nuclear weapons still exist in various states of readiness controlled by perhaps decaying command and control systems. An unknown number of nuclear weapons exist in politically unstable regions with failing nation-states. A quarter of a century after the fall of the Soviet Union, we do not know how many small "suitcase" nuclear weapons still exist. Complacency about the nuclear threat may decrease vigilance and so increase the likelihood of their eventual use. The challenge, however, is not to better control what cannot be controlled indefinitely, but rather to eliminate nuclear weapons altogether as Henry Kissinger, William Perry, Robert McNamara, Lee Butler, and many others have urged. This will require broadening the concept of national security and rethinking the legal basis for the nation-state that dates back to the Treaty of Westphalia in 1648. Difficult as that may be, it pales in comparison with what is at stake: everything humans have ever fought for, cherished, accomplished, and dreamed. The very possession of nuclear weapons has placed us at the edge of an abyss beyond our comprehension.[6]

Second, the vital signs of life on the planet are almost everywhere in rapid decline.[7] Climate change is the most ominous of these symptoms, but there are other signs that the planet is ailing. Human demands on air, waters, soils, forests, lands, and biota already exceed the Earth's carrying capacity by a large and growing margin. The result is a cascade of problems amplified by an increasing population and economic growth. Simply put, there are more of us with higher expectations and more stuff moving around the planet faster than ever before. The longer the economic and demographic juggernaut continues, the more difficult the transition to a manageable and humane outcome will be. Population growth and the extractive, high-throughput economy will end, either by our choice—hopefully in a manner that is fair, decent, durable, resilient, creative, and peaceful—or in catastrophe. But one way or another it will end.[8]

Nuclear Armageddon and ecological ruin work at different rates but either could destroy civilization. They are also inseparable. Nuclear war becomes more likely with climate changes that lead to a hungrier, thirstier, poorer, and more desperate world. And rapid climatic change is more likely in a world of competing, short-sighted nation-states inept in the arts of compromise and foresight.

A third self-induced and largely ignored threat to civilization exists at the periphery of our awareness: a drive for technological innovation propelled by our infatuation with novelty and the lure of large profits. Technology has become a virtual religion: we are so mesmerized by the new digital world in which we can communicate, fabricate, and fantasize apparently without limit that we can scarcely imagine any other. Premonitions of the darker side of technology put forward by writers and thinkers from Mary Shelley to Bill Joy, Nicholas Carr, Jaron Lanier, and Nick Bostrom, like those of Cassandra in mythology, have been mostly dismissed as Luddism or drowned in the intoxication that seems to accompany sophisticated technological wizardry. The trend is toward a world of total surveillance and mass manipulation, one that would have astonished even Orwell and Huxley. But that may be the least of it.[9]

Decades ago, MIT mathematician Norbert Wiener feared that humans might be displaced by machines that became smarter than their human creators. In 1948 he wrote, "If we move in the direction of making machines which learn and whose behavior is modified by experience we must face the fact that every degree of independence we give the machine is a degree of possible defiance of our wishes. The genie in the bottle will not willingly go back in the bottle, *nor have we any reason to expect them to be well disposed to us.*" Wiener, in short, was not optimistic that smart robots and machines could be controlled. Neither is engineer Daniel Crevier, who concludes in his history of artificial intelligence, "Machines will eventually excel us in intelligence, and it will become impossible for us to pull the plug on them . . . their arrival will threaten the very existence of human life." The main battles of the twenty-first century, he argues, will not be fought over environment or poverty, but "whether we or they—our silicon challengers—control the future of the earth."[10]

In 2000, Bill Joy, founder of Sun Microsystems, reached similar conclusions and called for a moratorium on any technologies, including artificial intelligence, that might acquire the ability to self-replicate. His warning, like the misgivings of Wiener and Crevier, has been ignored.[11]

On the horizon, perhaps sooner than later, is what mathematician I. J. Good once called an "intelligence explosion," that is, the creation of machines perhaps a thousand times more intelligent than humans. But the intelligence of these machines would be uninhibited by human corporeality, upbringing, culture, or morality. They would have no mother, father, siblings, or playmates. They would not have gazed at the stars in wonder or learned empathy from a pet dog. In other words, superintelligent machines would be alien to anything we presently know and quite possibly anything we could understand. They would be able to reprogram themselves, strategize, and set their own goals—goals that, one can presume, would prioritize their own survival and growth. As James Barrat explains: "We humans have never bargained with something that's superintelligent before. Nor have we bargained with *any* nonbiological creature . . . it won't have feelings. It won't have our mammalian origins, our long brain-building childhood, or our instinctive nurturing . . . It probably won't care about you any more than your toaster does." In Bostrom's words, we are "like small children playing with a bomb. Such is the mismatch between the power of our plaything and the immaturity of our conduct. Dealing with superintelligence with any sentience is a challenge for which we are not prepared and never will be."[12]

In fact, we will have only one chance to get this right and that is before "it" is born. After the dawn of sentient superintelligence, the situation will be beyond human control forever, and we have no good reason to believe that our creatures will be "well disposed to us." Despite the unknowns and terminal risks, the race to create superintelligence is being rushed along in artificial intelligence laboratories and in companies such as Apple and Google. The winner will presumably control much of the world for a brief moment until it, too, is displaced. At one level the development process resembles other competitions such as arms races between nation-states or the battle for market share between corporations. Beyond our foresight, these races are conducted without brakes and without the capacity to limit the ensuing effects. The situation is out of control, but not yet necessarily beyond control. "Sooner or later," however, in James Martin's words, "humanity has to learn how to control technology and avoid what is too dangerous, just as we must control a teenager if he is learning to drive a Ferrari."[13]

Our future may or may not resemble that portrayed in movies like *Terminator* and *The Matrix*, but given the momentum of the technology and the money to be made, it is difficult to give plausible reasons why it wouldn't. Superintelligent machines will not be cute, affable, empathetic, and servile like R2D2 of the *Star Wars* movies. They are far more likely to be cold, calculating, alien, totally other, and fully absorbed in their own agendas, which may well be unfathomable to us and hidden.[14] In the words of George Zarkadakis, "AI systems have the potential to become the ultimate controllers of everything." It is, he states, "a technology like no other."[15]

The challenges posed by nuclear weapons, climate change and ecological deterioration, and artificial intelligence have much in common. First, they are not anomalies, but rather the manifestation of the logic inherent to the systems in which they are embedded. Each will require permanent vigilance and long-term governance of a kind that we can scarcely imagine. More than simply problems, each is a dilemma in which defensible values of security, prosperity, justice, freedom, humanism, and progress clash. None of the three can be resolved within the paradigms and worldview in which they were created. Each is global, effectively beyond the capacity of any nation-state to solve unilaterally. The three are inextricably bound together as different facets of a single civilizational crisis with roots in the scientific revolution of the seventeenth century. They are sustained by an economy dominated by the imperatives of violence, speed, and growth, and justified and upheld by elaborate fraud. Together they represent complex, systems-level problems that cannot be remedied merely by changing their coefficients—the rates by which things get better or worse.

If humans are not merely to survive, but to thrive, the structures and their underlying assumptions that generated these lethal problems must be changed. Each has a threshold or trigger point that we cannot safely cross, and cannot know in advance. After the first detonation of a nuclear weapon, the start of runaway climate change, or the dawn of machine superintelligence, all the things we might have done "if only we'd known" are the stuff of epitaphs. Although none of the three has a solution, as we customarily use that word, it is possible that they can be managed long enough so that generations living in a wiser and less frenetic age will wonder why we risked so much for so little and why it took so long for us to see better possibilities.[16]

In the best of circumstances, there will be no quick resolution to problems posed by nuclear weapons, a warming climate, superintelligence, and their interactions. That means that we will have to create stable and effective systems of governance and management that will work on these problems over long periods of time and on a global scale. Only competent governments can do, and sooner or later must do, such things as eliminate nuclear weapons, tax carbon-based fuels, protect human dignity, improve economic fairness, counter sectarian violence, provide disaster relief, manage the global commons, and translate an informed public will into reasonable and enforceable policy and law.[17]

No one knows whether democratic societies can develop the capacity and stamina to master the complex challenges of the long emergency. There are many reasons why democracy is said to be a bad bet, most based on allegations that the public is incompetent, stupid, self-centered, easily manipulated, hopelessly divided, fearful, shortsighted, inconsistent, illogical, and lacking interest in or a capacity for the hard work of self-governance. The historical record does indeed show that democratic citizenries are often inept and indifferent. Yet, as Winston Churchill famously noted, the record for all other forms of governance is even worse. For better or worse, our best hope is to resuscitate democracy; improve civic education; promote ecological literacy; extend democracy throughout the economy, corporations, and society; evict all private money from the political process; improve education at all levels; reform public communication; improve representation; extend and guarantee voting rights; and shorten political campaigns.[18]

So how might political change of such magnitude begin? Given the formidable power and wealth defending the status quo, and the underwhelming record of utopian schemes, ideologically driven revolutions, and large-scale planned transformations, the only rational course may best begin with a steady withdrawal from dependence on large and unaccountable institutions and organizations and a commensurate rebuilding of the foundations of society from the grassroots. If so, the shift to sustainability and resilience would grow out of initiatives led by neighborhoods, transition towns, forward-looking cities, agile companies, better investors, and imaginatively led organizations of all kinds—with the goal that the positive changes will cascade upward to a regional, national, and global scale in ways that we can only dimly imagine.

In short, we have conjured at least three ways to terminate or disable civilization if not end the career of *Homo sapiens* altogether. We can destroy civilization in a sudden nuclear spasm devastating cities and covering the planet with a blanket of radioactivity. We could trigger runaway climate change that would destroy the ecological conditions that permitted humankind to flourish. Or we could create sentient and uncontrollable machines that would likely evict us. Some combination of the three is also plausible. But in one way or another all the important questions of our time are merely footnotes to the overriding issue of our tenure on Earth.

Of these three, I have chosen to focus on the threat posed by climate change. It is the most urgent of the three and also offers the greatest leverage to improve the human prospect. Rapid climate change will destabilize the international system and make nuclear war more likely. On a planet destroyed by nuclear weapons or rendered inhospitable to civilization, the threat posed by artificial intelligence becomes a moot issue. On the other hand, virtually all humans contribute—in widely varying degrees—to the problem of climate change. But we can choose to move away from fossil fuels to renewable sources of energy as well. To do that, however, we must "come to our senses" as Václav Havel puts it, changing our priorities, behaviors, goals, institutions, and lots more. In that clearer light we may come to appreciate all that we stand to lose and change our hearts, minds, politics, and economy accordingly.

THE CHALLENGES OF SUSTAINABILITY

two

> Man's nature being what it is, the destiny of the human species
> is to choose a truly great but brief, not a long and dull career.

—Nicholas Georgescu-Roegen

> Global warming is itself a sum of lost causes.

—Andreas Malm

A famous cartoon by Sydney Harris in a 2006 issue of *American Scientist* shows two scientists facing an equation on a blackboard. The key part of the equation reads, "Then a miracle happens." The second scientist deadpans: "I think you should be more explicit here in Step Two." Almost every proposal to improve the human future has a "then a miracle occurs" aspect followed by a skeptic's admonition to be more specific. But after thousands of books, reports, and articles written over many decades, we are still long on diagnosis and even prescription, but very short on how.

It is clear, however, that all plausible scenarios leading to a decent human future will require a combination of extraordinary citizen activism, dynamic and empowered democratic institutions, brilliant inventions, and creative, strong political leadership. The emergence of global grassroots climate activism led by 350.org and dozens of other organizations is the most encouraging trend of the past decade. The rapid deployment of solar technologies, too, is occurring much faster than expected. But it is difficult to believe that we will make a successful transition to a decent future without addressing the wider issues implied by "sustainability," a term that was introduced into the public dialogue with three books: *World Conservation Strategy* (1980), Lester Brown's *Building a Sustainable Society* (1981), and *Our Common Future* (1987). As Jeremy Caradonna points out, the word sustainabil-

14

ity goes beyond environmentalism and is "as much social, political, and economic as it is environmental history." Sustainability is the study of complex systems and the "relationships between society, economy, and the natural world." Political scientist Leslie Thiele writes that the goal of sustainability is to avoid undermining "the conditions that allow civilization to flourish within a supportive web of life." Brendon Larson similarly proposes: "sustainability involves our seeking a future in which the basic needs of humans are met, but without impairing (or destroying) the natural systems and species that support us."[1] The term has its limits and its critics: it is and will likely remain contested, partly because it is too firmly rooted in Western science and philosophy, and partly because we do not yet know what we must do to build a global, fair, peaceful, and durable world. But I believe that sustainability will not be achieved through managing the planet, as some propose, but rather by learning to manage ourselves within its limits. We aren't smart enough to do the former and must acquire the wisdom to do the latter.[2]

Opinions about our future range widely. Author Diane Ackerman, for example, writes, "We are dreamsmiths and wonder-workers. What a marvel we've become, a species with planetwide powers and breathtaking gifts . . . [who] can become Earth-restorers and Earth-guardians." Jane Jacobs, by contrast, believes that we may be "rushing headlong into a Dark Age," a time of "mass amnesia" that becomes "permanent and profound." Such times are "horrible ordeals," she writes, and "unsalvageable if stabilizing forces themselves become ruined and irrelevant." We may place our faith in technology to make our way through the peril, but culture is more than gadgets. It is instead composed of overlapping groups, such as institutions, organizations, and families; practices, including rituals, play, and ceremonies; and beliefs, which encompass our expectations, paradigms, hopes, and fears. As philosopher Clive Hamilton writes: "Clinging to hopefulness becomes a means of forestalling the truth. Sooner or later we must respond and that means allowing ourselves to enter a phase of desolation and hopelessness, in short, to grieve . . . bringing our inner experience into conformity with the new external reality will for many be a long and painful emotional journey." Our lives will be harder, but he hopes for a "resurgence of resourcefulness and selflessness," as well as "vigorous political engagement aimed at collectively building democracies that can ensure the best defenses against a more hostile climate."[3]

Scientist James Lovelock believes that it is already too late to prevent a catastrophic warming of the Earth, and foresees a remnant of humanity being forced into air-conditioned cities and "somehow merg[ing] with their electronic creations in a larger-scale endosymbiosis." We would no longer be human, but instead half-human, half-computer creatures like those described by Ray Kurzweil, on the other side of the "singularity" when the power of technology will become irreversibly life-altering for humans. By contrast, Paul Kingsnorth and his Dark Mountain colleagues, who have little faith in advanced technology, believe that it's too late to rescue civilization. The reasons they give are plausible, but they also could be a self-fulfilling prophecy, negating chances for better outcomes. There are many other views on the human prospect, but these four are representative of opinions that range from optimistic faith in technology and human creativity to total despair. Somewhere in the middle are the hopeful who believe that against all odds we will rise to the occasion. Despite their differences, all alert observers can agree that the situation is unprecedented, dire, and that the time required to make the necessary changes is very short.[4]

The transition to a more durable and just global order, accordingly, will be long and difficult for at least two reasons. The first and most obvious is the anarchical international system, which is rendered even more fractious by large differences in wealth and the centrifugal effects of politics, religion, ethnicity, and culture. Whatever its particularities of power and structure, the new world order must be inclusive and fair, with the means to resolve difficult conflicts. But it cannot be an empire administered by the United States.

There is another reason. The cumulative effects of population growth and fossil-fuel-powered economic expansion have destabilized and depleted the ecosphere with long-term consequences. We can no longer count on the relatively benign climate conditions that prevailed during the past ten thousand years. Rather, for a long time to come we must contend with the effects of the still-growing accumulation of heat-trapping gases in the atmosphere—effects that include rising temperatures, longer and more intense droughts, larger storms and cyclonic events, rising sea levels, acidifying oceans, and the loss of many species, as well as the social, economic, and political effects of life in a more threadbare, hungrier, hotter, and less climatically stable world. Every political, social, and economic problem will be magnified by climate conditions that will worsen for a long time to come. We do ourselves and our children no favor to overlook the duration of the journey and challenges ahead for fear of alarming the somnambulant or discomfiting the ideologues on either the Right or the Left.

In our predicament, the word solutions should be used with care. With enough money and time it should be possible to power the United States (and the world) with renewable energy, but only if we also greatly increase energy efficiency and improve the design of our cities and infrastructure. The global agreement reached in Paris at COP-21 is a significant, if tardy, step in that direction. But this does not mean that we will have solved the climate problem because of the long persistence of carbon dioxide (CO_2) in the atmosphere that will affect weather for centuries to come. And there are no practical solutions for species extinction or ocean acidification. In other words, there are no easy and quick ways to get out of the hole we've been digging for nearly two centuries.

There are, however, practical ways to minimize the eventual severity of the crisis and perhaps shorten its duration. These include increasing energy efficiency, rapid deployment of solar power, and increasing the terrestrial uptake of carbon through the improved management of forests, rangeland, and farmland. Yet none of these is a silver-bullet solution for the larger problem. As Australian climate scientist Tim Flannery explains: "If we put aside seaweed farming, with its stupendous potential but great difficulties in realization, the following optimistic scenario is within the bounds of possibility.

Forestry and soil carbon might together sequester a gigatonne of carbon per year, and biochar a similar amount. Direct air capture and silicate rocks might capture another gigatonne between them, and carbon-negative cement and carbon negative plastics another gigatonne. That's four gigatonnes of carbon per year, or around 15 gigatonnes of CO_2—just one quarter of current global emissions and still below the 18 gigatonnes that the combined US academies found we'd need to draw down to reduce atmospheric CO_2 by one part per million per year."[5]

But these management changes presume that various governments and landowners will have the wherewithal to better manage carbon on hundreds of millions of acres of farms, rangeland, and forests worldwide; that carbon uptake can be improved in varying ecologies under hotter conditions and through extremes of drought and severe flooding; and that carbon, once sequestered, will stay put. Much the same can be said of proposals for carbon capture and storage from coal-fired power plants. These would raise the cost of electricity and require transporting and storing massive amounts of carbon (roughly the same volume as all of the oil presently flowing through the economy) in suitable geologic formations, all in the hope it would not cause earthquakes, or leak back into the atmosphere, and could be done at a price deemed affordable. Proposals for using less energy by improving efficiency, or by end users "doing without," require a culture capable of separating the important from the ephemeral. But such things come as afterthoughts if they come at all.[6]

In short, the issues are contentious, the numbers are daunting, the scale is global, the time is short, and the requisite economic, social, and governmental changes are monumental. But this is no time for resignation or despair. Instead we need clarity, courage, and well-considered actions at many levels, all of which must be executed with sustained intelligence and a sense of urgency.

Our national history and psychology predispose us to assume that technology can fix virtually every problem, and adequate responses to the challenges ahead will certainly include better technology. But technological innovation can cause as many problems as it fixes, particularly when the issues are complex, systemic, and long-term. Dealing with the complex issues of climate change will be rather like solving a quadratic equation in mathematics in which each part of the equation must be solved separately, in sequence, to reach the right answer. In other words, we must learn the art and science of solving multiple problems without causing new ones.

To reach Montana, whatever that journey may come to mean, we will have to reckon with four interrelated problems beginning with those pertaining to our "intellectual emphasis, loyalties, affections, and convictions." The future of humankind depends on enough people changing very soon their feelings, thoughts, and behavior. All religions, as well as all parents, teachers, and every educational institution, will need to take the lead in this effort, by cultivating an affinity and unwavering attachment to life, along with an articulate, ecologically literate worldview that together inform attitudes, opinions, and behavior and so, in turn, lead to larger structural changes in governance and how we measure progress. The long emergency will require stable, effective, and truly democratic governments that protect natural systems and human rights alike, including those of future generations. For many reasons, including the deeply entrenched power of money and media, the necessary changes in our political economy will have to start at the local and regional levels and work upward until the sheer mass of small changes worldwide transforms the larger structures of civilization.

One large obstacle to that transformation began with a bargain made long ago with the giant companies that promised to make our lives affluent, convenient, easy, and fun. But they also made our world fragile, polluted, violent, undemocratic, insecure, and unsustainable. Moreover, they corrupted our very thoughts and words. As philosopher Alberto Manguel writes, "The 'happy language' of consumerism, of publicity, and of political slogans is employed to communicate short, simple messages representing nothing, aimed at convincing, never at opening an exchange, never at allowing in-depth exploration. [Its purpose is] to sell everything to everyone and, above all, boast . . . the virtues of the quick and easy . . . [our language] is desiccated to the point that its moral will always be one that satisfies our most egotistical desires." He continues, "The mercantile structure that we have built as the driving engine of our society is as perfect as those other imaginary constructs, and as lethal. We have given it the command to reach a goal, to render financial profit at all cost; we have forgotten to inscribe in its memory the *caveat:* except at the cost of our lives."[7]

George Orwell would not have been surprised. The corruption of language, he noticed, precedes the hollowing out of civilization, after which things predictably come undone. But it may be even worse than Orwell thought. The cumulative effects of increasingly sophisticated advertising (which reportedly bombards each of us with an average of five thousand messages a day) may have retarded our capacity to respond to abstract dangers like climate change. Sigmund Freud's nephew, Edward Bernays, was among the first to harness the emerging science of mass psychology and his uncle Sigmund's insight into human behavior. The aim was to create a new kind of person: a more dependable and dependent consumer. In Bernays's words, the goal was to "engineer consent" by addictive consumption in order to keep the masses quiescent and make money for their masters. "The conscious and intelligent manipulation of the organized habits and opinions of the masses," he wrote, "is an important element in democratic society." There was madness and purpose in the mission that aimed to disconnect us from reality and our money without revealing the fine print. Advertisers understood that humans are not so much rational creatures as they are very good rationalizers. Accordingly they learned how to mine the depths of the id and to exploit our vulnerabilities—particularly those of the most susceptible. Accordingly, children and adolescents are targeted with the intent of making them lifelong consumers and brand loyalists. The long-term consequences, for the climate among other things, are considerable. As Clive Hamilton notes, during "the last long boom the marketeers planted a poison pill deep within affluent society—a generation of children consciously moulded into hyper-consumers." They are the victims of a large percentage of the reported $500 billion spent for advertising worldwide each year. It is no accident that the first word spoken by many infants is a commercial brand name. Young children reportedly can identify more brand names than native plants and animals in their backyard. They are tethered to electronic devices that accompany them through childhood and adolescence. Early on they learn to want much and so learn little about restraint and self-discipline—or, for that matter, their backyards. That such words sound quaint and archaic is no small indicator of Bernays's success.[8]

It is time to undo that bargain and take back the power we assigned to anonymous agents who are loyal only to the goal of increasing their market share. It is time once again to declare our independence from arbitrary, capricious, and dangerous powers that threaten our survival.

There is a long history of rebellion against arbitrary authority of feudal barons, ecclesiastical overlords, overbearing royalty, and stone-hearted enslavers of all kinds. The origins of liberty are found in the earliest philosophical conversations about justice, mercy, obligation, and decency and grounded in the belief that each life has unique value and deserves to be protected. The chains are no longer of steel but they bind even more tightly. Aldous Huxley's *Brave New World* was a premonition that we could be manipulated and seduced into choosing our bondage and even come to prefer it over freedom.

The transition to a fair, decent, democratic, zero-carbon, and sustainable society will require transferring power to competent and thoughtful citizen-agents. It will require a public that is able and willing to think and do for itself and that is thereby less dependent for its information and provisioning on sources that it has little reason to trust. This is the work at the intersection of ecology, politics, ethics, and the new economics. It is emerging in the art and science of ecological design that is transforming architecture, engineering, agriculture, land use, forestry, waste management, and urban planning. It is being carried out by all of those working to transform neighborhoods, businesses, organizations, farms, towns, and cities. They are the vanguard of a better world.

three

RESILIENCE

> I tend to favor an economic system based on the maximum possible
> plurality of many decentralized, structurally varied, and
> preferably small enterprises that respect the specific nature of
> different localities and different traditions and that resist the
> pressures of uniformity by maintaining a plurality of modes of
> ownership and economic decision-making.
>
> —Václav Havel

A Marine Corps friend of mine defines resilience as the ability to "take a gut punch and come back swinging." More formally it is said to be the capacity to maintain core functions and values in the face of outside disturbance. Either way, the concept is elusive, a matter of more or less, not either-or. It includes the combination of slow, cumulative changes like soil erosion, loss of species, and acidification of oceans, as well as fast, "black swan" events such as the Fukushima disaster—all of which will create overlapping levels of unpredictable turbulence at various depths. The idea that we can improve resilience at scales ranging from cities to global civilization is becoming an important part of policy discussions, but if we are serious about resilience we will have to not only improve our capacity to act with foresight, but also develop the wherewithal to diagnose and remedy those deeper problems rooted in language and other paradigms, social structures, and our economy that undermine resilience in the first place.[1]

The theoretical underpinnings of the concept go back to the writings of C. S. Holling on the resilience of ecological systems and to metaphors drawn from the disciplines of systems theory, mathematics, and engineering. More recently, scholars such as Joseph Tainter, Thomas Homer-Dixon, and Jared Diamond have documented the histories of societies that collapsed for lack of foresight, competence, ecological intelligence, and environmental restraint. The concept of resilience is related to that of sustainability, but differs in at least one crucial respect. Sustainability implies a stable end state that can be achieved once and for all. Resilience, by contrast, is the capacity to make ongoing adjustments to changing political, economic, and ecological conditions. Its hallmarks are not just redundancy, adaptation, and flexibility, but also the foresight and good judgment to avoid the brawl in the first place.[2]

Yet we are inclined to whittle the discussion of resilience down to the simpler issue of technology, which has a bad habit of misbehaving in counterintuitive ways. While the use of better technology is certainly a large part of societal resilience, the definition of "better" is seldom obvious. The reason is that we do not simply choose to make and deploy single gadgets or innovations; instead we, rather unknowingly, select devices as parts of larger systems of technology, power, and wealth. The steel plow, for instance, represented not only the ingenuity of John Deere, but also an emerging but seldom acknowledged agro-industrial paradigm of total human domination of nature that spanned commodity markets, banks, federal crop insurance, grain elevators, long-distance transport, fossil-fuel dependence, chemical fertilizers and pesticides, crop subsidies, overproduction, mass obesity, soil erosion, polluted groundwater, loss of biological diversity, dead zones, and the concentrated political power of the farm lobby, which in turn represents oil companies, equipment manufacturers, chemical and seed companies, the U.S. Farm Bureau, commodity brokers, giant food companies, advertisers, and so forth. The upshot is a high-output, ecologically destructive, fossil-fuel-dependent, unsustainable, and brittle food system that wreaks havoc on the health of land, waters, and people alike. Farmers did not just buy John Deere plows; they bought into a system, and the resilience of that system had nothing to do with their choices.[3]

There is little in the modern world that is either resilient or sustainable. To the contrary, modernity has been about "the effecting of all things possible" as Francis Bacon once proposed in *The New*

Atlantis. Progress, that most loaded of words, is defined as increasing power, wealth, velocity, accumulation, and control over nature, and devil take the hindmost. But we are not so much like Bacon's intrepid, rational truth-seekers as like very clever apes tinkering in a warehouse filled with the most amazing but inexplicable things. The idea of resilience is largely alien to our cultural DNA. As a result, the drive toward globalization, more economic growth, faster communication, robotics, drones, nanotechnology, ever more interconnectedness, and so forth has created a global world without firebreaks or even fire departments. As the velocity of change increases, too, we have less and less time to reflect and mull things over. Without anyone intending it, we have created an increasingly fragile house of cards that hangs by the slenderest of ecological, energetic, social, and economic threads. But whatever else it is, it is no accident. Rather it is the logical working out of a system of ideas, beliefs, and "preanalytic assumptions" that are deeply embedded in the Western worldview. While there are some obvious things we can do at the national and international levels to improve resilience, such measures are no more than temporary stop-gaps that conceal deeper flaws rooted in our paradigms and worldviews. We are prone to tinker at the edges of the status quo and then are puzzled when things don't improve much and even larger disasters occur. My point is if we are serious about designing and building resilience, we will face a long and difficult process of not just rebuilding our hardware and infrastructure, but also rooting out those ideas hidden in our paradigms, language, political systems, economy, and education that undermine our resilience. Perhaps when we come to a fuller understanding of the discipline and restraint that sustainability and resilience will require of us, we may, like Thelma and Louise, prefer to go off the cliff in a blaze of glory. But if we decide otherwise, the conversation about resilience must advance from a focus on the coefficients of change to the structure of larger systems and ideas of human dominance—which is to say, from symptoms to root causes. Among other things, this will require returning to earlier conversations that go back to the likes of Lewis Mumford, Jane Jacobs, Herman Daly, John Ralston

Saul, and further back in time to Frederick Soddy, Karl Marx, John Ruskin, and John Stuart Mill and others who first noticed the cracks in the hard-shell presumptions of the modern project.[4]

The economist Nicholas Georgescu-Roegen's observation that in this world governed by the laws of thermodynamics, humans have a choice between a long and dull history and a brief but exciting one captures the essence of our situation. By some combination of destiny, accident, or choice we have taken the latter path, but we did not lack for warnings. Marlowe's *Doctor Faustus*, Mary Shelley's *Frankenstein*, Melville's *Moby Dick*, Conrad's *Heart of Darkness*, and science fiction films such as *The Matrix* are cautionary tales about the perils of overreaching, inattention, obsession, addictions, irresponsibility, and power run amuck. The sinking of the *Titanic* in 1912, too, provides perhaps an overused, but dramatic, metaphor for the hubris that afflicts our technologically driven society. We have been deaf to such warnings and proceeded in the faith that nature sets no traps for unwary societies. But any moderately well-informed high-school student could make a long list of plausible ways by which civilization could be crippled or end badly from the predictable consequences of its own actions.

As noted, at the top of such lists are the permanent threat of nuclear war and rapid climate change. Kids aren't dumb to the reality that in their lifetimes things will get a lot hotter. As noted earlier, if we don't change course we are heading for a warming of two degrees Celsius perhaps as early as midcentury and possibly of four to six degrees Celsius by 2100. Somewhere along that trajectory many things will come undone, starting with systems providing water and food, but eventually entire economies and political systems. Nearly everything on Earth behaves or works differently at higher temperatures. Ecologies collapse, forests burn, metals expand, concrete runways buckle, rivers dry up, cooling towers fail, and people curse, kill, and terrorize more easily. Climate deniers, of course, remain unmoved by science and the evidence before their eyes, but they are doomed to roughly the same status as, say, members of the Flat Earth Society. More serious problems arise from those who presumably know what lies ahead, but choose not to speak about it for fear of alarming the public. As a result of both denial and evasion there is a large chasm separating the science and current public discourse about planetary destabilization. But whether we face it or not, we will have to contend with the remorseless working of the big numbers that govern the biosphere. The "long emergency" ahead is not solvable in any way that we normally use that word. What we can and must do is to head off the worst of what lies ahead by making a rapid transition to energy efficiency, renewable energy, better design of cities and transportation, and changing the consumption/satisfaction ratio. Humans have never faced a more vexing and dire convergence of problems caused not as much by our failures as by our successes.

Historian Ronald Wright calls this a "progress trap" by which we attempt to solve the problems caused by progress with the same methods and mindset we used to progress. The heart of our predicament, he writes, is that "technology is addictive. Material progress creates problems that are—or seem to be—soluble only by further progress." It is an old story. As he explains, "Many of the great ruins that grace the deserts and jungles of the earth are monuments to progress traps, the headstones of civilizations which fell victim to their own success." The problem, Wright believes, is the inherent "human inability to foresee long-range consequences."[5]

We are preoccupied, in other words, with the here and now. What lies beyond our limited foresight is confusing and veiled and so we dawdle. Further, the often conflicting and competing value systems that direct our attention often draw us to one problem while making us blind to another. For instance, the headlines report the fast news from the latest scandal to the daily jiggles of stock market trends, whereas the brown color of the local river reports the slow movement of topsoil seaward. Even though soil erosion, literacy rates, and the disappearing ice sheets and glaciers say much more about our long-term prospects, the "fast-news" headlines are likely to capture most of our attention. Moreover, as these slow variables work over decades or centuries, baseline expectations and memories of better things shift downward to a new normal and we forget what once had been. Mesmerized by ever more powerful technology, we all too easily fail to notice these vulnerabilities silently multiplying and ramifying all around us.

Looking ahead even a few decades, the "progress trap" will lead to more difficult and unprecedented problems for which we are unprepared. Ray Kurzweil's proposal to merge humans with carbon-based intelligence to create something beyond—or below—humans seems like a monumental leap to me, but there is almost no public discussion about whether this is a desirable thing to do or not or even who has the right to make such decisions. We seem to be sleepwalking toward seismic and irrevocable changes in virtually everything we have until now regarded as fundamental to our humanity. When Bill Joy called for a moratorium on technologies with the capability to self-replicate, he might as well have been howling at the moon. We are unpracticed in foresight, precaution, and the discipline necessary to restrain and redirect our technological imperatives. The merest suggestion of caution with technology is the twenty-first-century version of heresy. We now have new and more powerful gods.[6]

A still more fundamental progress trap is inherent in the dynamism of a continually growing, energy- and resource-intensive, consumption-oriented market economy. The market economy has also attained a kind of divine status worshipped, worried over, and appeased with sacrificial offerings (for example, Detroit and Flint). There are good reasons to believe that the economy has already exceeded the carrying capacity of Earth. But the possibility that there are limits to growth or fundamental differences between quantitative and qualitative growth is still incomprehensible to most economists, corporate chiefs, bankers, financiers, managers of the economy, media talking heads, and all of the nabobs who gather to preen and be seen at their annual séance at Davos. Few of these or their hangers-on seem to notice the accumulation of ironies piling up all about them. Since the 1950s, for example, economies in developed countries expanded by three- to eightfold, but indicators of happiness didn't budge. Beyond some fairly low income level, we are no happier with more stuff than we are with less, even though rates of suicide, crime, and mental illness suggest that we are considerably more distracted and distraught. Accumulating wealth is increasingly offset by what John Ruskin once called "illth," in the form of pollution, climate change, unpaid social costs, and ugliness in its many guises. We are wealthier than ever, but the gap between the super wealthy and the rest of us continues to widen and the collateral effects of inequality and demoralization infiltrate every sector of modern society. Once we confidently presumed that our legacy was an unalloyed stream of benefits to our progeny, but the truth is that we have cast onto our descendants a lengthening shadow of biotic impoverishment, deforestation, acidic oceans, toxic pollution, and declining climate stability.[7]

What would a sustainable, fair, and resilient economy look like? What energy sources can dependably and benignly power it? How large an economy can be sustained within the Earth's limitations? How large an economy can fallible humans safely manage? What would it mean to give up our obsession to dominate nature? How will we distribute wealth? What would it mean to develop an economy for "Gross National Happiness"? How will we subtract from corporate balance sheets the $20 trillion of fossil fuels that cannot be burned safely? Who will decide such things? Questions like these have been shunted aside in the manic phase of industrialism, but if not for the well-being of all of the people and all of those to come, what is an economy for?

Such questions are first and foremost political, not economic. They have to do with how we provide food, energy, shelter, materials, transport, healthcare, and livelihood, and how the risks and benefits resulting from those choices are distributed. But these issues are often excluded from public deliberation and democratic control. From the beginning, the economic deck was stacked to protect wealth; individual rather than collective rights; and perversely, the rights of corporations as much or more than those of flesh and blood people. Further, it gives little or no protection to future generations even when their "life, liberty, and property" are put at risk because of the actions of the present generation. In short, the system is rigged to protect power and wealth and not to foresee or to forestall obvious risks such as a looming climate disaster. Our manner of governance seems incapable of reforming itself, let alone dealing proactively and constructively with the scale, scope, and duration of the perils ahead. Even at their best, it is debatable whether democratic societies are capable of exercising the foresight and precaution necessary to make resilience a priority in difficult circumstances.[8]

And so we get to the nub of the issue: the founding ideals of America had to do with equality, liberty, and justice, but these have always competed with other values embedded in the "American dream," which are mostly about the freedom of individuals to get rich. Early on, we were, in historian Walter McDougall's view, a nation of hustlers and deal-makers, celebrated now as "job creators." Pursuit of the American dream led to the indiscriminate exploitation of wildlife, water, soils, forests, minerals, and people. Our laws, regulations, taxes, and subsidies were designed to accelerate economic expansion and to make it easy for the lucky ones to make lots of money. At the same time we made it much harder than it had to be for minorities, Native Americans, the underprivileged, women, workers, unions, immigrants, the poor, and now, increasingly, the middle class. We've made it harder to exercise economic and technological restraint, foresight, and precaution even as the scale and scope of risks have become global and the damages irrevocable.

Do we have the right stuff for resilience? No smart gambler would bet on us. It's late in the game; there are more than seven billion of us on our way to maybe eleven billion and we've loaded the dice against ourselves by filling the atmosphere with carbon, sharply reducing the planet's biological diversity, acidifying the oceans, and spreading toxins and trash everywhere. Canadian biologist John Livingston, like Lynn Margulis, once described humans as "a rogue primate," and any sentient intergalactic review panel for *Homo sapiens* would certainly agree. But that is by no means all that we are. We also have the capacity for compassion, foresight, care, tolerance, ingenuity, creativity, decency, and, I believe, resilience. Given the lateness of the hour and the gravity of our situation, what's to be done?[9]

"Resilience," according to Donella Meadows, "arises from a rich structure of many feedback loops that can work in different ways to restore a system even after a large perturbation." Some of the first steps to improve the resilience of the United States are obvious. The engineering principles and technology needed for a more resilient electrical grid, for example, are well understood. A resilient power system would be distributed among many renewable energy sources. It would be highly efficient, carbon neutral, and organized around interlinked "smart" micro-grids that feature two-way communication between the grid and end-users. Energy prices would be based on the full lifecycle costs of energy, including its externalities. Consequently, it would use a fraction of the energy we presently use while providing higher-quality service.[10]

The principles of resilient urban design are also well known. In Eric Klinenberg's words, resilient urban areas consist of communities with "sidewalks, stores, restaurants, and organizations that bring people into contact with friends and neighbors." Healthy neighborhoods have many people watching the streets, as Jane Jacobs once noted, as well as many overlapping connections among churches, businesses, civic organizations, schools, and colleges. More resilient communities are pedestrian- and biker-friendly and their housing, schools, jobs, theaters, clubs, coffee shops, health facilities, and other attractions exist in close proximity. They have multiple and interconnected layers of "social capital," a rather clunky phrase describing competent, caring, and engaged citizens who work and play together and understand the importance of common wealth. Urban communities that are working to improve their resilience recycle wastes, minimize their carbon footprints, grow by in-fill, and are stitched together by pedestrian walkways, bike trails, and dependable, clean, safe, and affordable light-rail systems. Resilient cities will also have a growing percentage of locally owned businesses and community-generated wealth that stay put to create still more prosperity.[11]

At the national level, resilient economies are diverse, with redundant supply chains and few monopolies. They prioritize public good over private accumulation. Barry Lynn argues, however, that the Western economic model is moving in the opposite direction, so that we are "depend[ing] intimately on a system that we are making ever more interactively complex and tightly coupled. At the same time, we are abandoning redundancy, abandoning close management attention, abandoning closed monocultural safety systems." In the realm of national policy, resilience will require a larger definition of security than heretofore. We spend trillions for defense against often exaggerated external military and terrorist threats, while ignoring self-generated dangers that have jeopardized our economic livelihoods as well as access to food, energy, clean water, shelter, physical safety, and health care. Policy analyst Patrick Doherty proposes a "grand strategy" that connects policy and market demand for smarter growth with strategic investments that build resilient energy infrastructure and agricultural systems. Mark Mykleby and Wayne Porter, former staff members at the U.S. Joint Chiefs of Staff, suggest a "national strategic narrative, making sustainability the new standard for national and foreign policy and reconnect[ing] our political conversations with our highest values."[12]

In short, we do not lack for ways to improve the resilience of our infrastructure and our capacity to adapt and foresee coming challenges. But these are only the first steps toward resilience. Andrew Zolli warns, "None of these is a permanent solution, and none roots out the underlying problems they address." Moreover, our increasingly complex technical "solutions" may cause more problems than they solve. Financial risk analyst Nicholas Taleb puts it this way:

> Man-made complex systems tend to develop cascades and runaway chains of reactions that decrease, even eliminate, predictability and cause outsized events. So the modern world may be increasing in technological knowledge, but, paradoxically, it is making things a lot more unpredictable ... we are victims to a new disease ... neomania, that makes us [call] Black Swan–vulnerable systems— "progress."[13]

In Taleb's view, we are increasingly vulnerable to more and more severe black swan events as a result of increasing complexity, interdependence, and globalization.[14]

I conclude this chapter with four observations. The first is that it is impossible to make an unsustainable system resilient. Sooner or later the careless exploitation of land, water, forests, biota, and people will lead to disaffection, overshoot, and collapse. There are many variations on the theme, but the point stands. No system can be made resilient or durable on the ruins of natural systems or on the backs of exploited people. Design dictates destiny, but not in a direct and predictable way. Consequently in order to improve resilience, we will have to remedy the systemic flaws that have rendered our future increasingly precarious.

Second, the challenge of improving resilience must begin by reforming those structures of governance and political processes by which we decide issues of war and peace, taxation, education, research and development, healthcare, economy, environmental quality, and the basic issues of fairness. The political reformation, I think, must begin in the United States and it may be in ways somewhat reminiscent of the revolution we led in the years 1776 to 1790. In Al Gore's words, "The decline of U.S. democracy has degraded its capacity for clear collective thinking, led to a series of remarkably poor policy decisions on crucially significant issues, and left the global community rudderless." Corporations and markets do many good things, but seldom without rules, structures, oversight, enforcement, and the countervailing power of government. Our inaction in the face of climate destabilization is rooted in failures of regulation, politics, foresight, and leadership—failures that are attributable to the corrupting power of money that infects governments and the political process at every level. As a result, a small group of oligarchs hold our common future hostage.[15]

There are deeper structural issues as well. As Nicolas Berggruen observes, "the faster, wealthier, more connected, and more complex our scientific and technological civilization, the less intelligent our governance of it has become." His solution is to involve "citizens in matters of their competence while fostering legitimacy and consent for delegated authority at higher levels of complexity." The path toward resilience, in other words, will require a substantial upgrading of our collective capacities of foresight, coordination, and enforcement as well as greater fairness within and between countries and generations. In Berggruen's view, the key to good governance requires constraints on consumerism and "institutionalized feedback arrangements that favor the long-term and counter the ethos of immediate gratification." In the same vein, policy expert Leon Fuerth proposes reforming the Executive Office of the President to build "anticipatory governance . . . a systems-based approach for enabling governance to cope with accelerating, complex forms of change." These new strategies would require no heroic leaps, only the development of rational procedures of planning and policy development. But both would require a smarter citizenry and governing elites, untethered to big money, who understand systems, ecology, and the importance of the longterm, and moreover are motivated to act for the common good.[16]

Third, there are no purely national solutions to systemic problems of fragility. In an interdependent world, we will have to evolve institutions, laws, procedures, and "habits of heart" that make resilience the default at both the local and regional levels, even as we develop formal institutions, non-governmental organizations, and networks at the global scale. In fact, an efflorescence of civic capacity is emerging in diverse ways: from "slow" movements (food, money, cities), to organizations tracking carbon emissions of corporations, to women planting trees in Kenya and grandmothers installing solar panels in the Himalayas, to transition towns transitioning, and not the least to the growing role of elders in tempering our adolescent enthusiasms.

Fourth, as important as better technology is to a more resilient future, real solutions will also require improvements in our behavior and institutions: the rediscovery of big ideas, traditions, techniques, design strategies, and, even those quaint and mostly forgotten qualities of wisdom and humility in an age much enamored of self-promotion, surface appearances, and busy with trivialities. We are caught in a trap of our own making. If we are to escape the worst of it, we will have to extricate ourselves from dangerous aspects of our own unleavened cleverness and wean ourselves from the faith that more of the same will somehow work differently this time.[17]

One final note. Conferences on the subject of resilience might best be held in places like Detroit or Easter Island where there are ruins that remind us of our fallibility. Perhaps they might also begin with a reading of a prescient work like Shelley's sonnet, "Ozymandias." But such confabs are almost always convened in fairyland places like Aspen, Davos, or Paris, or in expensive hotels in Washington, D.C. amid the trappings of power, wealth, and aggrandizement where the well-dressed and expensively coiffured speak assuredly of endless opportunities in the comforting faith that a policy adjustment here or better technology there will suffice. For all of their elegance, power, and influence, such gatherings are not likely to advance the cause of resilience very much.

There were rumors of unfathomable things, and because
we couldn't fathom them we failed to believe them, until
we had no choice and it was too late.

—Nicole Krauss

In 1897, Swedish scientist Svante Arrhenius published his calculations on the impacts of burning fossil fuels on the Earth's climate. His measurements showed that a doubling of CO_2 in the atmosphere would raise Earth's temperature by five to six degrees Celsius, which is still regarded as ballpark accurate. Since Arrhenius's time, the temperature of the Earth has increased by slightly more than one degree Celsius (1.7 degrees Fahrenheit) with maybe half as much or more on the way—even if we were to meet the goals set at the COP-21 conference in December 2015. The years from 2011 through 2015 were the warmest yet recorded. Temperatures are increasing faster at the poles and on land than in oceans. Despite a rapidly growing deployment of renewable energy resources, a two-degree Celsius increase in temperature by 2050 is thought to be probable (but not inevitable). If this does occur, the planet will be warmer than it has been at any other time over the past several million years. The last time the Earth was that warm was 120,000 years ago, when its sea levels were thirteen to twenty-six feet higher than today. Unless our course changes dramatically, the warming will not stop there; a further increase of four to six degrees Celsius by 2100 is also considered possible. At some point along that trajectory the result would be "absolute chaos, the end of civilization."[1]

Even in the most optimistic scenario, the Earth's temperature will continue to rise for a long time to come. Today we are experiencing weather extremes and ecological changes that were not thought likely until the Earth's carbon dioxide levels had risen to approximately 450 parts per million (ppm) or higher. And the twenty- to thirty-year lag between the carbon dioxide coming out of our smokestacks, tailpipes, and soils and the weather means that the extremes we are experiencing now are the result of what we did several decades ago. Because CO_2 remains in the atmosphere for a long time, we have committed generations to come to centuries or millennia of rising temperatures, rising sea levels flooding coastal cities, rapid ecological changes, loss of a large fraction of our biological diversity, and collateral effects including famine, violence, political instability, economic decline, and psychological trauma.[2]

Further, a slowly declining fraction of CO_2 will remain in the atmosphere for thousands of years. In the words of the Intergovernmental Panel on Climate Change: "A large fraction of climate change is largely irreversible on human time scales unless net anthropogenic carbon dioxide emissions were strongly negative over a sustained period." Or as geophysicist David Archer puts it: "The climatic impacts of releasing fossil fuel CO_2 to the atmosphere will last longer than Stonehenge, longer than time capsules, longer than nuclear waste, far longer than the age of human civilization so far." Twenty-five percent of CO_2 released today will "still be affecting the climate one thousand years from now . . . about ten percent will still be affecting climate in one hundred thousand years." So even were we to stop emitting greenhouse gases today, the Earth's temperature and sea levels would continue to rise for a thousand years or longer. This is emphatically not simply an environmental issue but rather a civilizational crisis that crosses every political boundary on Earth and will disrupt every sector of every society for centuries and millennia to come. "One of the most robust predictions that can be made," in Elizabeth Kolbert's words, "is that [climate change] will send millions—perhaps tens or hundreds of millions—of people in search of new homes." In the near to medium term the most obvious effects of climate change will be rising levels of disease, crime, violence, inflation, economic chaos, unemployment, large storms, and international conflict. And in an interconnected world, "disaster cannot be cordoned off"—one calamity will lead to another.[3]

The faintly pleasant, anodyne term "global warming" does not begin to describe the magnitude, severity, and longevity of the reality ahead. "Planetary destabilization" is far more accurate. Virtually everything that we presently take for granted will change, much to our disadvantage. Because of past emissions, our future will be marked by larger storms, longer and more severe droughts, changing ecologies, rising ocean levels, loss of species, record heat waves, along with collateral effects including famine, water shortages, economic turmoil, domestic violence, political instability, and resource wars. The pace of change is much faster than was predicted even a few years ago. Consequently we are now experiencing weather extremes and other changes that were not expected until mid-century or later.[4]

Still unclear are how certain "wild cards" will amplify the warming effect. Methane (CH_4), for one, is a shorter-lived but far more potent greenhouse gas that is released as the temperatures of boreal soils and sea water rise. Beyond some unknown temperature threshold, massive releases of methane from soils and methane clathrates in shallow seas would likely occur. It is the climate equivalent of a time bomb, and the fuse is lit.[5]

There are other positive feedback loops, both fast and slow, that will increasingly amplify and accelerate climate changes. One, already evident, is the change in the albedo (reflectivity) of polar icecaps to darker seawater that absorbs more heat, thereby accelerating warming. In other words, the climate system is "non-linear." As climate scientist Wallace Broecker once said, "Climate is an angry beast and we are poking at it with sticks." The volume and rate of carbon dioxide emissions are unprecedented in the geological record. As a result, biological systems have no time to adjust.[6]

As a policy problem, climate destabilization is particularly "wicked." For policy wonks "wicked" issues are those that are "ill-defined, complex, systemic, and purportedly unsolvable," entailing "multiple constituencies with conflicting agendas and . . . comprised of seemingly unrelated, yet interdependent elements, each of which manifest as problems in their own right, at multiple levels of scale." Sometimes wicked problems can be endured until the circumstances that created them change. Otherwise they can be only contained or perhaps managed. Wicked problems are the sort that discomfit our managerial presumptions, overwhelm our siloed understanding of how the world works, and undermine our economic and social order. The results of the wicked problem of climate destabilization in particular will exacerbate struggles to maintain international security because it will "likely lead to food and water shortages, pandemic disease, disputes over refugees and resources, and destruction by natural disaster in regions across the globe." Climate change is a "threat multiplier" that makes every existing security problem worse while adding entirely new ones such as masses of climate refugees crossing international borders.[7]

In short, the scientific evidence is clear: we are rapidly undermining the ecological and climate conditions of the Holocene that allowed humankind to flourish. In the presence of overwhelmingly obvious danger, we have so far failed to act in ways commensurate with the scale and projected duration of the crisis and the extent of death and suffering that further procrastination will cause. Many of the effects could have been avoided had we acted when the situation was clear decades ago. The worst effects are still presumed to be manageable if we avoid crossing the threshold of increasing the planet's temperature by 1.5 degrees Celsius, but they will become progressively less so beyond that number. The result of five decades of procrastination and denial of the scientific evidence was to make the human prospect much more precarious than it had to be.[8]

Nothing I've written so far will seem controversial to the 97 to 99 percent of scientists who study climate for a living and must abide by the rigors of peer review, fact, data, logic, and scientific evidence. At the simplest level, the science is straightforward: heat-trapping gases in the atmosphere trap heat in the form of incoming short-wave radiation from the sun. A lot of heat-trapping gases in the atmosphere will trap a great deal of heat by preventing the escape of long-wave thermal radiation back to space. The sciences involved include atmospheric physics, chemistry, and ecology, the interactions of which are considerably more complex. Disagreements, such as they are, occur at the thin edge of a very large consensus. And as a general rule the more we've learned about climate science and Earth systems, the better the future appears to be for jellyfish and cockroaches and the worse for *Homo sapiens*.

Given the overwhelming scientific consensus, it is odd that climate change has become so politically charged. The possible reasons for this are beyond the scope of this paper. Suffice it to say that much of the purported controversy has been conjured by those with a large stake in the approximately $3 trillion-a-year business of selling coal, oil, and natural gas and with an estimated $20 trillion (or more) worth of reserves yet to sell. Unable to resolve the science in their favor, their strategy has been to sow confusion, which has caused further delays and billions in profits. By comparison, the much less certain and far less portentous threat of terrorism has so far caused us to spend trillions of dollars without hesitation and with no end in sight.[9]

In sum, we are in the early years of a long-term and accelerating disruption of the Earth's systems that if allowed to go on much longer will be fatal to civilization. As Elizabeth Kolbert writes: "It may seem impossible to imagine that a technologically advanced society could choose, in essence, to destroy itself, but that is what we are now in the process of doing." The primary causes are the combustion of coal, oil, and natural gas as well as changes in land use that result in nine billion tons of carbon released each year to the atmosphere. Given the residence time of carbon in the atmosphere, there is no known technical solution to our self-induced crisis. But it is technically and economically feasible and would be thoroughly intelligent to make a rapid shift to a post-fossil-fuel society powered by efficiency and renewable energy. Carried out on a global scale, it is our best, fastest, and cheapest way to avert the worst that otherwise awaits.[10]

Why does it seem, for the moment at least, that the sheer magnitude of the risk has exceeded our capacity to comprehend and respond? One reason is that evolution has shaped human nature to be optimistic against long odds. This trait has served us well. After what appears to be irretrievable disaster, the survivors have persistently sucked it up and gotten back to the work at hand: breeding, hunting, gathering, farming, inventing, stealing, selling, trading, bamboozling each other, and stacking up rocks for temples and castles to be made into future ruins—future tourist sites—thereby providing income for their progeny. We are also well equipped to deal with dependably loathsome enemies like Adolf Hitler and Osama bin Laden and with immediate threats that are hairy, long-toothed, and fearsomely clawed. Honed in the plains of eastern Africa, the amygdala area of our brains, in the face of physical threats, quickly signals the adrenal system, which in turn gives instructions for "fight or flight." Early on we evolved as highly successful hunters and warriors adept at killing big animals and fending off predators and nasty neighbors. As a result, we have evolved to be better fighters, defenders, explorers, and conquerors than cooperators, conciliators, and peacemakers—roles that require more complex thinking and qualities like empathy. In the face of an oncoming tiger, for example, empathy was a bad idea; few empathizers survived to leave their genes in the pool. Even today, our heroes are still mostly the conquering sort, whose likenesses are displayed on horses and in military getup celebrated as pigeon garnished statuary in city parks. Schoolteachers, fire fighters, social workers, and those who clean up the mess and get the kids back to school seldom rate such status. In other words, our evolution has not kept pace with the circumstances we have created.[11]

When shown graphs of ocean acidification, species extinction, or the annual accumulation of CO_2 in the atmosphere, most Americans have shown little interest and until very recently not the slightest twitch of fear or apprehension. Signs of alertness—heart rate, breathing, adrenal flows, oxytocin levels, and the like—remain constant or even decline. Marinated in a culture that doesn't much value smartness, we are bored, even offended, by such facts. Conditioned by decades of commercial advertising, compulsive consumer self-indulgence, easy mobility, movies with happy endings, and the endless search for commercialized fun all made possible by artificially cheap energy, we are repelled by anything that demands intellectual effort or sacrifice for the common good, or that risks a bad outcome. So the typical response to being told the facts of climate destabilization is to dismiss them because they are too depressing, as if science must be rendered commensurate with one's preferred emotional state and perpetual fun were a Constitutional right. Heaven forbid that anything should ever be depressing even when it is. The possibility that we have brought on for ourselves and our descendants a long run of climate-change-driven catastrophes is as indigestible as a rock.[12]

Accordingly, those who persist in telling the truth about climate change are admonished to be more positive and talk only about the many opportunities in the green economy. There is certainly a case to be made for sunny optimism. Optimists are a lot more fun than pessimists. And sometimes they do defy the odds and snatch victories from the jaws of defeat. But the line between optimism and delusion is often fuzzy. Optimists often are wishful thinkers only because the alternatives are too painful to confront. In poet T. S. Eliot's words, "Human kind/Cannot bear very much reality." Sometimes we can't or won't handle the truth. In either case, it is possible that public lethargy in the face of rapid climate destabilization may be more cultural than evolutionary. Americans, in particular, seem to have made a fetish of optimism and positive thinking no matter what the science or the numbers. As Barbara Ehrenreich explains, "Americans [have] been working hard for decades to school themselves in the techniques of positive thinking, and these included the reflexive capacity for dismissing disturbing news."[13]

A third possibility is that attitudes and opinions about climate change reflect the distribution of power and wealth. The powerful prefer that some things remain hidden and that some dependencies be perpetual. This is an old story. In *Brothers Karamazov*, Fyodor Dostoyevsky, in the book's most famous chapter, has a fictional Grand Inquisitor say to a silent Christ:

> Oh, never, never can they feed themselves without us! No science will give them bread so long as they remain free. In the end they will lay their freedom at our feet, and say to us, "make us your slaves, but feed us."[14]

The Grand Inquisitor and the Church, in this story, have a good thing going and it all hinges on keeping the masses dependent, gullible, incompetent, fearful, and hungry. The same relationship may help to explain why much of the American public has been slow to respond to impending climate disaster. If one "follows the money," as any smart attorney would, the trail leads to the lairs of the oil, gas, and coal vendors who preside over a $3 trillion per year business. We now know that ExxonMobil executives knew by 1977 that the company's oil and gas would drive global warming, but they chose to fund a robust denial industry, and that the fossil-fuel vendors also own or influence much of the media and advertising industry, which has unsurprisingly ignored or downplayed the issue of climate change. Knowing this, the supposed mystery of public lethargy in the face of the largest threat humans have ever faced evaporates like water on an asphalt parking lot in Houston. Certainly there are other factors, but for the sake of simplicity, like Occam's famous razor, we might trim away the fluff and blame our inadequate response on the distribution of power and wealth in society, combined with garden-variety corporate greed that operates in a culture with certain propensities somewhat influenced by our evolutionary dispositions.[15]

As we look to the next generation of leaders for action on climate change, we should remember that real leaders cannot sound an uncertain trumpet or equivocate. They lead, rather, by clearly identifying a problem, setting forth practical goals, and inspiring people to follow or better yet, become leaders themselves. They help us think clearly, see possibilities, and motivate action. It works similarly at all levels.

Imagine, for example, Winston Churchill at the height of the Nazi blitzkrieg on London in 1940 saying to the British people over the BBC: "Hitler has given us a wonderful opportunity for urban renewal! We can finally build the city that Christopher Wren proposed after the Great London Fire of 1666." Instead, he summoned them to face the hard realities of war and offered only "blood, toil, tears, and sweat." They responded heroically. Similarly had Martin Luther King soft-pedaled the truth about lynchings, beatings, discrimination, and poverty in American culture, the "fierce urgency of now" would have been a lot less fierce, urgent, and memorable. Faced with life-threatening illnesses, people can and often do rise to a higher level of awareness and behavior. Confronted by disasters, people more often than not show a remarkable capacity to self-organize and cooperate. And told the truth, entire generations can rouse themselves to do what Thomas Berry calls their "Great Work."[16]

But there are notable differences. We can be roused to heroism more easily if the circumstances are dire and short-term: after a few years of sacrifice, even the hottest fires of wartime enthusiasm burn out. The effects of climate change, however, will persist for much longer. To preserve what could still be a decent future we will have to act quickly, yet to succeed we have to have the stamina for the long haul. So what should we say to a still skeptical public?

Mark Twain's advice was simply, "When in doubt tell the truth, it will confound your enemies and astound your friends." In our predicament, however, telling the truth raises many perplexities. Do we tell the whole truth or just a bit of it? Might it be doled out a little at a time like bitter medicine? Does the public need to know that climate destabilization will, at some unknown point, put everything they value at risk, including the lives of their children and grandchildren? Would they be paralyzed by such knowledge, dissolving into a puddle of fear and irresolution that would lead them simply to go shopping? Is it enough that most understand something about the reality of the problem without grasping its severity and longevity? Should we break down the problem into, say, steps that should be taken in the next decade or two to decarbonize the economy, which perhaps *is* a solvable problem? Or should we talk about "climate solutions" as if it were actually possible to (1) suck two centuries of accumulated carbon out of the atmosphere and thereby restabilize the climate, which is improbable, or (2) geoengineer the surface of the Earth and the atmosphere and incur all manner of ecological and political risks that cannot be known except after the experiment has already been run?[17]

In other words, should we seek a strategy that breaks a complex, wicked problem into smaller and more digestible pieces? If we focus only on the next few years or decades, global warming might be portrayed, for example, as an opportunity to create jobs and make lots of money by selling "climate solutions" in a growing green economy—unless, that is, you live in places already being flooded out, baked out, or blown out. Depending on how one sees it, this is either an evasion of the longer-term realities or a good way to buy time in which to develop a more comprehensive approach to the crisis. This strategy, too, has the great advantage of engaging people now in doing what can be done, albeit with the risks that they will not try hard because they will continue to believe the problem is easily solvable and that they will become disillusioned later when they learn the truth.

If the great mass of people is unreliable and fickle, should we take Plato's counsel and accept rule by elites? If so, exactly which secretive, self-appointed elites are up to the challenge of climate leadership? Should we entrust the financial wizards who collapsed the global economy in 2008 and haven't done much better since? Or maybe the neocon disciples of Leo Strauss who engineered the Iraqi War? Or perhaps the anti-elite libertarians who are eager to dismantle our capacity to act as a public body? How about the liberal crowd of climate activists that spins the facts of our situation to avoid alarm? Or should we lean on the best and brightest scientists whose tribe also gave us nuclear weapons, mutual-assured destruction, long-lived pesticides, and thalidomide—and infected African Americans with syphilis in the 1930s? The truth is that there is no one to save us; we will have to do it ourselves. We can no longer afford "delusional optimism" on one hand and despair on the other.[18] Instead our last, best hope lies in courage to hear the truth about climate destabilization and the will to defy the odds.

five

ECONOMY

'Tis the times' plague, when madmen lead the blind.

—William Shakespeare

The cost of a thing is the amount of what I will call life which is required to be exchanged for it, immediately or in the long run.

—Henry David Thoreau

It feeds, clothes, houses, entertains, transports, employs, invests, disinvests, and showers all manner of things on those with money. It also poisons and pollutes while making some rich and many others poor and some fat while others starve. It flattens mountains, destroys ecologies, acidifies oceans, destabilizes the climate, creates continental-sized gyres of trash in the oceans, and corrupts democracy. It is creating amazing new ways to displace humans in favor of robots and render our carbon-based minds obsolete in favor of those made of silicon. It provides a multitude of ways to communicate 24/7 while making it less likely that we can talk sensibly with our neighbors. It entertains and infantilizes, perhaps while "consuming itself." Both bane and blessing, it is the global capitalist economy. Three centuries in the making, it has grown into a world-straddling colossus that levels cultural differences from Shanghai to Madrid. It dominates our politics and news. A slight blip up or down in the market can cause mass euphoria or gloom. It is said to have begun in the ancient urge to "truck and barter" and in the turbulent winds of greed, envy, ambition, and fear. Sometimes, however, it unleashes more benign forces of creativity, innovation, and philanthropy. It is a vast and incomprehensible machinery comprising banks, financiers, investors, entrepreneurs, corporations, tax-dodgers, foundations, workers, child laborers, capital flows, government agencies, legislative committees, lobbyists, business schools, professional economists, TV savants, peddling advertisers, compulsive consumers, hidden networks of influence, black marketers, organized crime, cyber-thieves, drug lords, and trillions of dollars of investment capital that wash around the world every day in search of a tenth of a percent higher rate of return. Those left behind are part of a growing and ominous reservoir of discontent. Anyone who purports to understand the global capitalist economy does so, at best, only in part and mostly in hindsight. The various opinions and theories seldom penetrate below the surface, since deeper explanations must reckon with the ecosphere of which the economy is a subsystem, with the structure of class and privilege, and with the roots of human behavior all the way down to the ancient reptilian brain stem where goblins and ghouls lurk in the shadows. The result is a jerry-built economic system that lurches from boom to bust exhibiting the extraordinary madness of crowds that seizes on tulips in one time and smartphone apps in another.[1]

The theoretical foundations of modern economics were first described by Adam Smith in *The Wealth of Nations* (1776). But Smith had previously written a major work about the bonds of sympathy that hold societies together (*The Theory of Moral Sentiments*, 1759) and was working on a new edition of it when he died. In one he is said to have argued the importance of self-interest; in the other the importance of empathy and sympathy. Whatever one's opinion about what Smith really thought, subsequent generations of economists built the shambling edifice of economic theory on the foundations of self-interest not sympathy, individual desires not public interest, private wealth not commonwealth, and the present not the future. They also assumed insatiable wants, endless growth, infinite substitutions for resources as they become scarce, know-how over know-why, and the existence of a chimerical creature called "economic man" who maximizes the foggiest of foggy concepts called "utility," a particle in the social/psychological universe that has never been seen nor its tracks ever detected. One's utility is, therefore, whatever one believes it to be—including, I suppose, finding one's utility in banishing the concept of utility forever. The mainstream neoclassical version of economic theory ignores the laws of thermodynamics, presumes away limits imposed by ecology, devalues nature as mere resources, assumes a model of human nature that would not pass muster among mentally healthy psychologists, advocates behavior repugnant to any ethical ethicist, defies the requirements for adequate safety margins that elsewhere inform good engineering, mixes its description of economic behavior with its prescription for proper behavior, and confuses rationality with our bottomless capacity to rationalize almost anything including the most abhorrent, depraved, ridiculous, idiotic, improbable, and hair-brained exhibits in the catalog of human behaviors. With some notable exceptions such as Arthur Pigou, John Kenneth Galbraith, Kenneth Boulding, Robert Heilbroner, and Herman Daly, economic theory from Smith to the present works best if the questions are small, the accounting is narrow, the time horizons are short, and its practitioners are true believers. Yet it towers virtually unassailable against legions of critics, including some of its own most distinguished economists. Its mathematical models, comprehensible only to the suitably inducted, are otherwise virtually bulletproof against logic, data, biophysical reality, actual experience, and the consequences flowing from the profession's own modest predictive performance, which some believe to be on a par with those of interpreters of chicken entrails, palm readers, and TV weatherpersons. Yet more than any other body of thought economics has come to define us as self-maximizing economic automatons independent of society, not as thoughtful, attentive citizens, or as dutiful community members, caring parents, spiritual creatures, or as the beneficiary of the labors of earlier generations and ancestors of those yet to be born. It has taken shallow to a whole new depth. By the circular logic of self-interest it purports to explain the saintly behavior of Mother Teresa and the bizarre megalomania of, say, Donald Trump. It is said to explain everything. Nobel Prize–winning economist Gary Becker, for example, once announced that

> the economic approach is a comprehensive one that is applicable to all human behavior, be it behavior involving money prices or

imputed shadow prices, repeated or infrequent decisions, large or minor decisions, emotional or mechanical ends, rich or poor persons, men or women, adults or children, brilliant or stupid persons, patients or therapists, businessmen or politicians, teachers or students.[2]

Professor Becker's revelation comes as a considerable relief to those still vainly laboring in now obsolete fields such as history, philosophy, psychology, politics, sociology, linguistics, theology, and literature. They may now stand down, thereby saving financially stressed institutions large amounts of money and thousands of students the aggravation of becoming "well rounded" when flatness is more efficient for career success. Perhaps the newly unemployed professors will have time to go bowling together. But I digress.

It comes, then, as no surprise that all problems are presumed to be economic and so can be solved only by economic solutions that mostly have to do with selling more of something unneeded to people who can't afford it in order to increase the wealth of those already overburdened with too much—all of which further accelerates the treadmill. This process, called "neoliberalism," is merely turbocharged capitalism. It is not an innocent theory. As David Harvey explains: "Neoliberalism has, in short, become hegemonic as a mode of discourse. It has pervasive effects on ways of thought to the point where it has become incorporated into the common-sense way many of us interpret, live in, and understand the world." Its main accomplishment has been "to redistribute, rather than to generate, wealth and income." One might suspect that this was the intention all along. As Maggie Thatcher once said, "Economics is the method; the object is to change the heart and soul." And in contrast to all who believed that the study of economics was merely to elucidate a complicated field of human behavior, it did change hearts, minds, and souls, creating a sizeable cadre for whom the abstraction of the market has become scriptural.[3]

More progressive economists admit many of the flaws of economics in theory and practice but believe nonetheless that the capitalist economy can be remodeled as "green" capitalism without the exertion of examining and re-forming its foundational assumptions. A smarter, circular, solar-powered economy would offset the negative aspects of institutionalized greed, perpetual growth, and consumerism. A few smart adjustments at the margin, a policy shift here, better technology plus a change in taxation, and violà . . . sustainability! Most of the proposed changes would be an improvement of the necessary but insufficient sort; perhaps the economic version of what theologian Dietrich Bonhoeffer once called "cheap grace." Alas a sterner and less forgiving reality is rapidly forcing us to rethink the basic premises of economics. We need to change the economy to accord more closely with: (1) the way the world works as a physical system, (2) the basic rights of present and future generations, and (3) the obligations that go with being a "plain member and citizen" of the community of life, and do so for reasons of our own self-preservation if we can find no other. The considerable power of greed without guardrails can drive GDP into the stratosphere and generate technological miracles for a time, but it can also drive civilization predicated on our worst traits over a cliff. And economic doctrine, in Robert and Edward Skidelsky's words, "has allowed the profession to maintain an attitude of cheerful indifference to the facts of human psychology."[4]

Economic theory, however, did not develop in a vacuum. Rather it coevolved with business practice and particularly in interaction with increasingly powerful concentrations of capital called corporations. Both theory and practice were thereby shaped by political and judicial decisions that placed the corporation at the center of our burgeoning economic life. We can trace the origins of the corporation to the Dutch East India Company and its British cousin, the East India Company. Both existed as licensed monopolies acting in lieu of the state. Subsequent opinions vary widely on the evolution of this hybrid but now dominant creature. On the one hand, corporations "allowed society to use markets much more effectively," while on the other, they "locked in a societal focus on the market providing material, not social goods." They can be credited greatly for the material abundance Americans enjoy and they can be equally blamed, among other things, for overflowing landfills. They can be credited for our mobility and thermal comfort and blamed for climate change. They are both a blessing and a curse.[5]

Even in its early stages unrestrained corporate power aroused considerable fears. Thomas Jefferson, for one, had premonitions of what lay ahead, writing in an 1816 letter: "I hope we shall crush in its birth the aristocracy of our monied corporations which dare already to challenge our government to a trial of strength and bid defiance to the laws of our country." In 1864, Abraham Lincoln similarly wrote to his friend William Elkins that "corporations have been enthroned and an era of corruption in high places will follow, and the money power of the country will endeavor to prolong its reign by working upon the prejudices of the people until all wealth is aggregated in a few hands and the republic is destroyed." Premonitions and fears, however, did not matter much against the gathering forces of economic evolution. By the middle of the nineteenth century, in legal scholar Morton Horwitz's words,

> the legal system had been reshaped to the advantage of men of commerce and industry and the expense of farmers, workers, consumers and other less powerful groups within the society. Not only had the law come to establish legal doctrines that maintained the new distribution of economic and political power, but, wherever it could, it actively promoted a legal redistribution of wealth against the weakest groups in the society.[6]

The political and legal foundations for the industrial-extractive economy were in place. What remained was to position the abstraction of the corporate form as a legal entity on a par with or superior to actual flesh and blood people who bleed, curse, cry, sing, suffer, live, and die. In one of the oddest legal developments in the history of U.S. jurisprudence, the U.S. Supreme Court, in *Santa Clara County v. Southern Pacific Railroad* (1886), supposedly decided that the corporation was, indeed, a legal person under the terms of the Fourteenth Amendment. The actual history, however, is less straightforward. The purported decision, incidental to the case itself, was not in fact rendered by the Court, but written in the headnotes to the text of the decision by the clerk to the Court who, as fate would have it, also had close ties to several railroad companies. "There was no history, logic, or reason given to support that view" in the words of Supreme Court Associate Justice William O. Douglas. Mr. J. C. Bancroft Davis, clerk to the Court, however, had his own reasons and logic and so took it upon himself to bestow a handsome windfall on his erstwhile benefactors in the railroad business by announcing in the headnotes that the Court presumed that corporations were included under the due process provisions of the Fourteenth Amendment, which were otherwise intended to address the urgent and more obvious needs of "persons of previous servitude" whose personhood was rather more palpable. With a pliability exhibited only by circus contortionists and lawyers, the protections of due process for former slaves became applicable to an abstraction. Corporations, transmogrified into persons, were invested with the protections of free speech and the right to lobby and contribute to political campaigns. Testifying to the enduring power of human credulity as Phineas T. Barnum once observed, we have come to accept as normal and beneficial the grotesque "personhood" of the entities that provision us, believing that they are dedicated to our benefit and enjoyment. In time that personhood would be greatly amplified by other and even more mischievous decisions similarly perched on Mr. Bancroft's precarious legal wizardry. The beneficence would someday spread further, until it permitted corporations to patent life, to propagandize as a tax deductible form of free speech, to lie with exuberant impunity, and more recently in *Citizens United*, to purchase the institutions and paraphernalia of democracy itself. The resulting co-habitation of corporations with democracy qualifies neither as a marriage nor perhaps quite yet as abduction, but the dinner conversations are strained.[7]

In the wake of the economic collapse of 2008, the inestimable Alan Greenspan, chair of the Federal Reserve, famously found a flaw in his own economic thinking. He suffered "shocked disbelief" and was "very distressed" to discover that hungry foxes make indifferent guardians of financial chicken coops. Greenspan's aha moment was rather like that of a commercial pilot flying at 35,000 feet who suddenly discovers something interesting about the law of gravity that he'd previously overlooked. Neither the passengers nor the chickens would be greatly amused. For Greenspan a stroll through the streets of Detroit or Youngstown a few years earlier might have illuminated theoretical flaws not otherwise visible from the commanding heights atop the Federal Reserve. Of economic theory generally, economist Paul Krugman says that the previous three decades of macroeconomics has been "spectacularly useless at best, and positively harmful at worst." Even so it is easy to lose sight of how strange and recent the market fetish is. In the late Tony Judt's words:

> Much of what appears "natural" today dates from the 1980s: the obsession with wealth creation, the cult of privatization and the private sector, the growing disparities of rich and poor. And above all, the rhetoric which accompanies these: uncritical admiration for unfettered markets, disdain for the public sector, the delusion of endless growth.[8]

It being assumed that this is just the way things are in the best of all possible economies, many became oblivious to the increasingly threadbare public estate and their increasingly precarious private circumstances and in that state of mind happily embarked on the political equivalent of a lengthy Australian walkabout. In the meantime others busied themselves over "the last three decades methodically unravelling and destabilizing . . . the dikes [public institutions] laboriously set in place by our predecessors." Yet, as Judt asks, are we so sure "that there are no floods to come?" The obvious answer is that floods will come, and we have good reasons to believe that they will be larger than those before, but the public capacity to foresee, forestall, or at least to repair the damage may be only a distant memory of a competent civic culture that once existed. How did this happen?[9]

One answer has to do with the diminished expectations and performance of those who profess to lead. On the ramparts of the economy one may still behold the captains of finance and business strangely oblivious, in this age of information and science, to basic truths about how the Earth works as a physical system, as well as to why such knowledge, whether deployed or not, bears importantly on their management of the commerce of the nation, and to why the quaint notion of civic obligation should still rouse their curiosity about possible connections between the two. In the academy, where they were once diligent and perhaps idealistic young scholars, we witness a Byzantine system of training and certification that could have only anesthetized and deadened their otherwise burgeoning idealism. The same instructional mechanism is further embedded in a vast system of "knowledge production" that may range between useless and harmful. Perhaps it is a small consolation to laborers in the vineyard of economics that the medical profession itself may not have crossed the breakeven point—the point where by one estimate it made a positive contribution to the health of those it purported to serve—until sometime early in the twentieth century. Before that time, a visit to the doctor actually lowered one's chances of survival. The word is "iatrogenic," meaning physician-induced illness.

To its legions of critics, it is unclear whether economics has graduated from its iatrogenic phase or not. Leaving that question aside, however, it is a useful time to examine the fundamentals of the discipline and what the great economist Joseph Schumpeter once called its "pre-analytic assumptions," or what is now more casually called the unseen "elephants in the room" that are so taken for granted as to remain unnoticed, unremarked, and therefore unstudied and unaccounted for.[10]

Actually there is a herd of elephants in the seminar room, including many of the ideas and assumptions that gave rise to the industrial economy. Things once presumed true are known to be less true than once believed, or even altogether false. Most important of these is the belief that fossil fuels are either inexhaustible or can be replaced by something better. Imaginatively deployed faith, stubborn naiveté, and garden-variety corruption have caused those in power to sit on their hands. Consequently, in the half century after the first oil embargo we still do not have a coherent energy policy, which means at this late hour we still have no binding de jure climate policy, and the question of what will power the next economy remains unanswered. Whatever the sources, they will not likely have the same return on investment or energy density that oil gave us early in the twentieth century.[11] They will also come with costs and consequences yet to be revealed. Other "elephants" include how we remove from the ledger books—assuming that we do—the portion of fossil fuels that we cannot burn without laying waste to the human prospect on one hand or causing undue hardship on the other.[12]

Our energy choices will affect others, including the food system. Americans pay comparatively less for food than everyone else on Earth. But the true cost is hidden beneath multiple subsidies for energy, land, and capital that keep the costs of food artificially low. As a consequence one calorie of food on the plate requires between eleven and seventy calories of fossil energy to grow, transport, process, refrigerate, and cook. Further, climate instability will cause increasing havoc on farms due to drought, heat, flooding, and novel ecological conditions. We should not presume a perpetual flow of foods from distant farms at prices we can afford and at a volume that we need. Neither should anyone else. Similarly, the reality of growing water scarcity looms darkly over the future of the Southwest and in the Midwest, where agribusiness is earnestly draining the last half of the Ogallala Aquifer. But the prospects are even more dismal in many parts of the Middle East, Africa, and South Asia, areas that have been affected by both permanent desiccation and rising sea levels.[13]

Other, more technical, arcana in the arsenal of economics must also be recalibrated to different and more constraining realities. When the future was confidently thought to be provisioned by an unfailing cornucopia, investors could discount the effects of catastrophes thought to be no more likely to occur than, say, an asteroid falling on Wall Street during its few hours of work. In less beneficent and less predictable times, however, the rate at which such events are "discounted" back to "net present value" requires rethinking. Economists will long debate how the prospect of ill tidings should be considered in making long-term financial decisions. Truth be told, no amount of academic quibbling over the appropriate discount rate relative to scarcely imaginable events at some time in the indistinct future can change the reality that the economic implications of climate destabilization are beyond mere human reckoning. Moreover, such events are no less real or less consequential for being difficult to fathom and predict with tolerable precision. We are entering a period of extreme climate uncertainty that will stress our prevailing economic theories and business practices, which were fashioned for less demanding times.[14]

In short, Adam Smith and those dutifully following in his tracks innocently presumed broad and continuing progress, measured by material conditions in tons, acre feet, cubic feet, square footage, sales, and above all, profits. They could confidently expect that such progress would continue far into the future sustained by presumably inexhaustible supplies of fuels, minerals, timber, soils, and bounty from the oceans. They could also expect stable ecosystems and a full range of natural services including climate stability. These conditions were simply taken for granted, as well they should have been given the state of knowledge in their time and the much smaller scale of the population and the economy. Similarly, they were predisposed to assume the superiority of British and Western culture over all others. The proof was said to be in the pudding. They were empiricists, who having tasted the pudding, became optimists of the imperializing sort. Their ideas, culture, technologies, and economy were considered to be timeless, at least for a while.[15]

Economic theory followed in due course. Virtually every assumption of classical and neoclassical economics, preanalytic or otherwise, is an outgrowth of conditions existing at the dawn of the industrial world. Theory followed facts presumed to be immutable. Alas, reality has a habit of making fools of those who presume too much.

A great deal has changed. Against what appears to be an increasingly bleak horizon, questions arise. We might pause to ask, for example, what useful economic theory might better fit our different circumstances and offer useful guidance for the perplexed? The issue, of course, is not whether we have an economy or economic theory or even a discipline called "economics." Instead we need to know what kind of economy we have, its rules, and who qualifies—by virtue of what training and experience—as a helpful guide on matters pertaining to the economy and its appropriate niche in the wider field of human ecology. It is worth noting as well that somehow previous societies got along tolerably well without a specialized caste of economists. Moreover, some that we would deem primitive provided decent lives for their members without a priesthood of economists and without much that any respectable economics department would recognize as economic theory. In our far more complex world, however, economists can play a useful, if perhaps more muted, role than they presently do: possibly, as John Maynard Keynes once proposed, with a social standing roughly equivalent to that of dentists.[16]

It would be presumptuous, however, to say exactly what flag economists should rally around or the level of status they should rightly aspire to. Accordingly, I will eschew overreach and confine myself to what I think is obvious and noncontroversial and so risk violating the other pole of possible critical outrage, that of being tedious and boring. But in truth, the facts, ecological and economic, are well known, making it all the more odd that they have not been applied with dispatch as remedies toward making a better and more durable economy. I therefore suggest an economics and resulting business practices oriented around four well-known principles.

1. *The economy is a subsystem of the ecosphere* and is thereby bound by its limits and is subordinate to the bio-geo-chemical cycles, energy flows, and ecological functions that govern the Earth and the health of its constituent parts. But the relationship is entirely asymmetrical. The ecosphere has no need or affection for an unruly and ungrateful tenant and could proceed quite well through the subsequent billion years or so without the spindly-legged, big-brained, narcissistic, and perpetually delinquent upstart that proudly calls itself *Homo sapiens.*

The economy, in other words, must conform to the rules set by the larger system in which it is embedded or sooner or later it will cause its own destruction. The demands of the economy for resources and energy and for absorbing its waste products, including the approximately 100,000 or so chemicals and all of their various combinations, must not exceed what the larger system or its component ecosystems can provide in perpetuity. Further, the larger system is known to be "nonlinear"—that is, unpredictable and prone to sudden changes. To the alert and prudent, the possibility of nasty surprises would suggest the kind of precaution that keeps wide margins. For an accident-prone, juvenile species it would further suggest sobriety in our interventions in natural systems that we understand only imperfectly. Yet we seem to live in the faith, as biologist Robert Sinsheimer once put it, that nature lays no booby traps for unwary species.[17]

The point is that no subsystem can grow indefinitely within its larger system without destroying itself and its host. "Perpetual growth," someone once said, "is the ideology of the cancer cell." Nonetheless, faith in endless economic growth and continual material expansion on a finite and "full" planet persists as the keystone myth of our time. The slightest mention of a "steady-state" economy, proposed by John Stuart Mill in 1848, typically triggers an avalanche of ridicule and the superior disdain of the pedigreed. *The Limits to Growth*, published in 1972, is still widely dismissed as irretrievably errant, which it is not. Beneath the surface of credentialed incredulity I suspect that there is something more at work that critics are loath to confess. As long as economies expand we can defer difficult and contentious issues such as the fair distribution of wealth, the effects of employment, and the effects of the things we make and sell to each other. When growth, measured as increasing quantity, ends—either because it collides with the finiteness of the Earth or because we can no longer manage the rising complexity resulting from the massive scale of the growth economy—we will have to reckon with the many problems associated with the highly skewed distribution of wealth and its relation to domestic tranquility.[18]

We will also have to reckon with the fact that we are not nearly as rich as we presume ourselves to be. The prices we have been paying seldom reflect the full costs of things purchased. Instead we have been offloading "external" costs onto others in some other place or at some other time. In Juliet Schor's words: "When we finally and fully tally up the costs of fishery collapse, soil erosion, desertification, wildfires, loss of tropical forests, toxic releases, and a mass extinction of species, the price tag will loom large." Some costs, such as climate destabilization, soil erosion, biological extinctions, and human exploitation at some point are simply beyond reckoning. Had we been on a pay-as-you-go plan would we have industrialized differently? Or not at all? Or might we have "developed" in more modest ways?[19]

Such is the value of hindsight. Meanwhile, however, all of the economic gauges, dials, and indicators record indiscriminate expansion as if it could go on forever. But it won't and the reasons are well known. The four cubic miles of primeval goo dug up and burned each year to miraculously power the industrial economy is also its Achilles' heel. In contrast to all previous civilizations, ours is powered by the exploitation of a one-time endowment of fossil fuels—ancient sunlight heated, compressed, and rendered dense and portable by millions of years of geology. The vast scaffolding of the modern growth economy was erected on the flimsy faith that this endowment from the carboniferous age was inexhaustible and further that it could be burned with impunity—that is, without cost to the health of people, land, wildlife, and waters. Americans accordingly increased their energy consumption 150-fold between 1850 and 1970. Virtually everything we make, use, eat, wear, build, refrigerate, light, and transport depends on burning fossil fuels. But fossil fuels did much more than energize our industrial age; they also changed our experience of the world. Night became day. Distance shrank. Time was compressed. Fossil fuels became lodged in our muscle memory and psyche as the thrill of power and in our sense of space and time as the exhilaration of speed. Combustion changed how we think and what we think about. In some ways it made us a dumber people unable to think clearly about limits and the work of repair, among other things. As historian Bob Johnson writes: "Having inadvertently skipped around the energetic limits to the solar economy, Americans became subsequently vaccinated against talk of ecological constraints."[20]

It was, however, a Faustian bargain and the devil will have his due. The carbon we have moved from where geology had safely stored it to the atmosphere will cause global havoc for a long time to come. But the deal has been coming undone for other reasons as well. Energy analyst Richard Heinberg, for example, has tracked the energy it takes to find, extract, process, and transport energy, or the energy return on investment (EROI). Predictably, as our fossil-fuel reserves have declined, the energy return on the energy invested has declined as well. A century ago a hundred units of energy could be extracted for one unit spent on exploration, drilling, mining, refining, and transport. Today, by contrast, the EROI for conventional oil is about 25 to 1 and falling. It will continue to fall as energy deposits are discovered farther out, deeper down, and often in places where people don't much like us. Without an outbreak of intelligence we will be stuck to a tar baby of unsolvable, expensive, and interminable conflicts for decades to come. In other words, we blew through the easy stuff and now must spend more and more to reach, process, and fight for our access to what's left. Will we run out of oil, gas, and coal? Not likely, but we have already exhausted the cheap and easily accessible reserves and what remains will be bitterly fought over. As Bob Johnson writes: "In coming into our energy inheritance, we behaved badly . . . the modern self is in crisis, and so getting sober and waking up to where we have been during that spree might not be the worst place to start."[21]

Could we power our present civilization by hyper-efficiency and sunlight in its various forms? Optimists like Amory Lovins believe it is possible if we are smart enough to manage better technologies with consistently flawless competence. I want very much to believe this to be true. Ozzie Zehner, on the contrary, writes: "Little convincing evidence supports the fantasy that alternative-energy technologies could equitably fulfill our current energy consumption, let alone an even larger human population living at higher standards of living." But even if we could, why would we want to? Why would we choose to maintain a standard of living based on waste, overconsumption, ecological ruin, fantasy, multiple addictions, exploitation of the powerless, growing inequality, and most assuredly, perpetual conflicts in distant places? In other words, why would we choose to do efficiently and with renewable energy many things that should not be done in the first place? Instead it may be possible for renewable energy to power a less frenetic society rebuilt around efficiency, social justice, frugality, and the long term—a society that, as Gandhi once said, meets everyone's needs but not their greeds, which are mostly generated by fears of one kind or another. Whether that society would be capitalist or something else altogether remains to be seen.[22]

2. *The economy is a means, not an end.* The purpose of a good economy is to provide and fairly distribute sustenance for living such as food, water, shelter, and healthcare. It would also provide for basic services including education and infrastructure for transportation and communication. It would create the means by which everyone could reliably earn a decent livelihood by doing good and necessary work and thereby increase their competence and capabilities. Beyond basic needs, it would further encourage the arts, beauty, kindness, solace, and conviviality for everyone, not just the well-off. In addition, in a decent economy prices would tell the truth by including all costs and externalities.

But a good economy would not grow merely for its own sake. Nor would it be driven by the creation of artificial needs by advertising that preys on children and ruthlessly exploits our needs for status, affection, and connection. A few decades ago Robert Kennedy famously made the same point:

> Gross National Product counts air pollution and cigarette advertising, and ambulances to clear our highways of carnage. It counts special

locks for our doors and the jails for the people who break them. It counts the destruction of the redwood and the loss of our natural wonder in chaotic sprawl. It counts napalm and counts nuclear warheads and armored cars for the police to fight the riots in our cities. It counts Whitman's rifle and Speck's knife, and the television programs which glorify violence in order to sell toys to our children. Yet the gross national product does not allow for the health of our children, the quality of their education or the joy of their play. It does not include the beauty of our poetry or the strength of our marriages, the intelligence of our public debate or the integrity of our public officials. It measures neither our wit nor our courage, neither our wisdom nor our learning, neither our compassion nor our devotion to our country, it measures everything in short, except that which makes life worthwhile.[23]

Simon Kuznets, the author of our present system of national accounts by which we measure economic performance, less poetically noted that "the welfare of a nation can, therefore, scarcely be inferred from a measure of national income." On the flip side, some argue that growth directly or indirectly improves the lives of people, and up to a point and in some ways it does. But economic growth conceals all manner of contradictions as well as a large number of goods and services that are deleterious to human well-being and to the larger human prospect. Capitalists, it is said, will sell anything for a profit, but some things such as climate stability, human health and dignity, sacred groves, children, and grandmothers should not be for sale at any price.[24]

A good economy would also facilitate that vague but important indicator of social health called happiness, which is notoriously difficult to define and measure. What we do know, however, is that the U.S. economy generates high rates of depression, autism, loneliness, violence, and addictions of various kinds, but not so much happiness and satisfaction. By reputable measures happiness in America peaked in the 1950s, when people had less stuff and there was far less of it to buy. The widening gap between things possessed and flatlined happiness is an otherwise inexplicable embarrassment to those who believe that these march in lockstep. If that were true, at about

9:00 a.m. on August 11, 1992, when the Mall of America first opened its doors, we would have crossed over into the state of perpetual Nirvanic bliss. But we didn't. Further, the ever-increasing gap says a great deal about the chasm between the economy's needs and those of flesh and blood human beings.[25]

3. *The economy must be nonviolent.* If the economy is to exist harmoniously within the ecosphere, it cannot do violence to the larger host system without fatal consequences to itself. Neither can a durable and decent economy employ violence in its various forms without harming not only the people it purports to serve but also the social and ecological conditions that underwrite their well-being. A bit of perspective may help. Our economic ideas originated in a culture with a long history of violence from religious crusades, inquisitions, imperialism, militarization, the making of—and perpetual threat to use—nuclear weapons, and continual wars over one thing or another. As a nation we grew rich from the proceeds of violence done to people brought here as slaves and to displaced Native Americans. Some benefited by doing great violence to landscapes, soils, forests, and wetlands. Violence was implicit in the extractive economy that prospered by wrenching wealth from the Earth and people alike. Violence was also present at the founding of science. Francis Bacon, the founder of the Royal Academy of Sciences in London, once described the scientific method as putting "nature on the rack and torturing her secrets out of her." Ever since, our motto has been, as William McDonough puts it, "If brute force doesn't work you're not using enough of it." The modern economy, in particular, depends a great deal on the research, manufacture, and selling of more efficient ways to kill. The defense budget, which includes the cost of wars, "security," surveillance, and eight hundred or more military bases worldwide, exceeds a trillion dollars. Much of the economy depends on that constant, unquestioned, and well-distributed flow of largesse to defense contractors. We glorify violence in our movies, advertisements, politics, and sports, and all too easily overlook the violence happening in our name in places like

Guantanamo, Abu Ghraib, and "black ops" sites where unnamed and unrepresented detainees are tortured for reasons that torture morality, logic, law, reason, bodies, and souls alike.

That we have become a violent, gun-toting society is no new insight. But violence also permeates the larger culture in less obvious ways. Our meat is raised mostly in animal gulags made efficient and profitable in direct proportion to the suffering inflicted on the confined animals. The application of machine-intensive, industrial farming methods has destroyed perhaps half of our topsoil in the United States. Farms, forests, and private lawns are managed mechanically and chemically, which is to say, violently. Modern medicine, too, applies industrial methods to the human body. After a century of promiscuous chemistry, every baby born in the United States arrives "pre-polluted" with several hundred chemicals delivered through the mother's umbilical cord. We have indeed, as Bacon advised in *The Great Instauration*, constrained and vexed nature: "forced [her] out of her natural state, and squeezed and moulded [her] . . . to the effecting of all things possible."[26]

Other economic and scientific possibilities beckon. British economist E. F. Schumacher writes: "Wisdom demands a new orientation of science and technology towards the organic, the gentle, the nonviolent, the elegant and beautiful." His proposal for "Buddhist economics" is premised on "simplicity and non-violence." Nonviolent methods for natural-systems farming and forestry are well known as being both practical and profitable. Similarly, biomimicry, the study of how nature makes virtually everything without combustion, toxic chemicals, pollution, and ecological ruin, can transform methods of manufacturing. Some of the most promising methods of avoiding disease and healing represent a merger of Eastern and Western practices and philosophies.[27]

4. *Politics are more basic than the economy.* In other words, the issues of the scale, purposes, and content of the economy are not first and foremost economic issues but political ones about "who gets what, when, and how." The "economy," accordingly, reflects the laws, regulations,

tax policy, government budgets, and political customs by which wealth is created and distributed. Karl Marx thought otherwise, arguing at considerable length and density that the political system is an outgrowth of the economy. With no small irony, many, if not most, neoclassical economists would agree that economics is more fundamental than politics and that economics mostly determines our political reality.

The relationship between the economic and political realms is certainly intimate and reciprocal, but if democracy really matters there can be no useful debate about their priority. The rules that govern the economy should be made public by properly elected representatives serving an informed electorate, not by an oligarchy meeting behind closed doors. The public should participate in decisions having to do with the distribution of wealth, risk, reward, and the sustainability of the entire human enterprise. They should be made aware of the social, political, and ecological consequences of economic and fiscal decisions, taxation, purchasing, and investment. To abdicate that responsibility and assign such decisions to the abstraction of the market, as economist Karl Polanyi once said, "would result in the demolition of society." It would also cause a deeper crisis as more and more people became caught in an inhuman system. In Václav Havel's words:

> A person who has been seduced by the consumer value system, whose identity is dissolved in an amalgam of the accoutrements of mass civilization, and who has no roots in the order of being, no sense of responsibility for anything higher than his or her own personal survival, is a *demoralized* person. The system depends on this demoralization, deepens it, is in fact a projection of it into society.[28]

The upshot is that most economic crises, recessions, depressions, and other perturbations are not accidents or anomalies, but rather the normal working out of the logic built into the rules of the economic system. Unrestrained and minimally regulated capitalism has led relentlessly to greater concentration of wealth, economic monopoly, and ever larger ecological, social, and political crises. But the system rules that inform the design, structure, and workings of the economy are seldom discussed publicly by economists in ways comprehensible to the lay public. In fact, "proficiency in obscure and difficult language," as John Kenneth Galbraith once noted, "may even enhance a man's professional standing." This may explain, in part, the public somnolence in the face of the widening chasm of wealth distribution and the lack of public accountability. In the United States, by one report, the richest twenty people have more wealth than the lowest-earning half of the population—that is, 152 million people. Worldwide, sixty-two people have more net wealth than the lowest-earning 3.6 billion people.[29]

Such facts would not have surprised Karl Marx who, with remarkable thoroughness, analyzed the dynamics of capitalism and its tendency to concentrate wealth in fewer and fewer hands. French economist Thomas Piketty, in similarly exhausting detail, has updated the trends of income and wealth, showing that "inequality reached its lowest ebb in the United States between 1950 and 1980" largely because of policies aimed to deal with the Depression and World War II. After 1980, however, inequality "exploded" mostly because of shifts in taxation and finance. Piketty is not a determinist as was Marx. Rather, he writes that "if we are to regain control of capitalism, we must bet everything on democracy." The upshot is that inequality eats away at the social fabric of democratic societies. As Richard Wilkinson and Kate Pickett show in *The Spirit Level*, virtually every bad social trend from crime to obesity is strongly linked to the unequal distribution of income and opportunity, risk and reward. The fact is that the global capitalist economy is trending toward greater concentration of wealth and is therefore increasingly prone to lurch from crisis to crisis and to foster a growing public disaffection. It is widely believed that the collapse of the Soviet Union was due to flaws that do not afflict capitalism. But in fact, communism and capitalism have many similarities, including their dependence on economic growth and an industrial paradigm rooted in a rationalistic philosophy that regards the world as so much dead material waiting to be transformed in ways suitable for human use and then discarded without consequence. Both ideologies worship at the same altar, even if they argue about who owns the church. In the larger sweep of history this is a Lilliputian-scale dispute. Certainly different in important respects, but the flaws in both systems bear a family resemblance, causing thinkers as different as Václav Havel and Jane Jacobs to conjecture that the collapse of capitalism cannot be far behind that of the Soviet Union.[30]

Others are not so sure. David Harvey, for one, writes that "capitalism will never fall on its own. It will have to be pushed. The accumulation of capital will never cease. It will have to be stopped. The capitalist class will never willingly surrender its power. It will have to be dispossessed." Maybe so, but others, like ecologist Howard Odum, argue that "general systems principles of energy, matter, and information are operating to force society into a different stage in a long-range cycle." Either way, the large numbers that drive historical trends—growing human discontent, declining energy return on investment, and the accumulation of unpaid social and ecological costs of our consumption—are converging toward a systems change.[31]

Fifty miles south of my home in Oberlin, Ohio, the large and thriving Amish community in Holmes County earns its keep mostly by farming, construction, and craft work. They travel in horsedrawn buggies with an effective range of eight miles and a hauling capacity of no more than several hundred pounds with a top downhill speed of twenty miles an hour. They farm with horses, pay in cash, do not accept Social Security, insure themselves, care for their elders, preserve their community against long odds, and do well in good years and bad. They do not attend school beyond the eighth grade, but they are known to be lifelong readers and are the largest users of the Holmes County Library. As farmers, they are, in Gene Logsdon's words, a "great embarrassment to American agribusiness" that presumes farms must be large, capital-intensive, specialized, subsidized, dependent on petrochemical "inputs," and forgiven their sins against soils, waters, ecologies, wildlife, public health, and posterity.[32]

For the Amish, farming is a family affair at the scale that a person can walk in several hours and situated in a supporting community for whom the word "neighbor" is a verb. The glue that holds the Amish community together originated with the Anabaptists four centuries ago and is a faith practiced daily and honed every Sunday in small congregations meeting in the homes of members. It is a tough-love community that upholds its own but also shuns those judged to be wayward. It permits its adolescents time to experience the temptations of the outer world, but gathers most of the temporarily wayward back into the fold. Amish culture, in Gene Logsdon's words:

> resists financial chaos . . . [and] fortifies individuals against their own frailty. The culture sanctifies the rural virtues that make good farming, or good work of any kind, possible: a prudent practice of ecology, moderation in financial and material ambition, frugality, attention to detail, good work habits, interdependence (neighborliness), and common sense.[33]

The Amish dress alike, simply and without show. They drive the same model buggy that, with proper repair, can last a lifetime. They follow neither fashion nor the latest trends. They mull over technological changes, like cell phones, for a long time before deciding whether to adopt or not. They are not good consumers and what they do purchase is mostly hardware and very practical. In other words, they control for the sin of pride and avoid the presumptions of large scale. They are pacifists and know how to forgive those who trespass.[34]

They are also shrewd businesspeople who do not put all of their eggs in one basket. A typical Amish farm will sell a dozen or more products throughout the year and will consistently make money when the conventional overcapitalized and less diversified ("English") farmers go broke. They make their own horsedrawn equipment in small shops that employ a dozen or two individuals and export to other farmers across North America. Like all farmers a century ago, before the tidal wave of mechanization and specialization, Amish farmers know how to do a great many things. They are decent mechanics, builders, plowmen, veterinarians, and helpful neighbors. A few, like David Kline in Holmes County, Ohio, are also good writers.

The Amish are not a perfect community. They are patriarchal, conservative, and their existence, when measured against the frenetic lives of the rest of America, may seem inexpressibly dull. They don't travel far or often. They refuse to watch television, tweet, google, or email. They do not exhibit themselves on Facebook. They do not build vast cities or make F-16s, or fly off into space. Neither do they make nuclear weapons, wage wars, acidify oceans, change the climate, and shortchange their progeny. They live on Earth with modesty, grace, and neighborliness. The horse sets the speed limits of their society, and their common culture, grounded in applied Christianity, fences in their personal and collective ambitions. They are an island in the vast sea of a modern world infatuated with novelty and stuck anxiously on an accelerating treadmill of change.

The Amish are also dismissed by the mainstream society as cute, quaint, and rustic, fit mostly to be set pieces in a charming landscape to entertain hordes of restless tourists taking a fall drive through "Amish country" in their SUVs. But we might also find them to be instructive on matters that perplex the non-Amish. In particular, they have, in their own way, solved problems that befuddle mainstream, Nobel Prize–winning, neoclassical economists. For example, the Amish economy is not particularly vulnerable to the booms and busts of the dominant global economy. Second, they keep the accounting wide to include collateral and otherwise nonmonetized costs. They refuse the seductions labeled "economies of scale" and thereby do not suffer the inefficiencies of "efficiency"—surely the most misleading word in the English language. They do not inflict their "externalities" on others downstream or downwind. Further, they have organized their society to maintain its small scale. When any community grows too large, a part moves out and forms another community elsewhere. No Amish community or church grows beyond the limits of people to know each other as neighbors. There are no mega Amish churches and so there is little self-righteousness and commercialized proselytizing. Accordingly, they do not suffer much from the loneliness and isolation characteristic of the wider culture. Third, the Amish have avoided the pitfalls of consumer addiction. Unlike the mainstream American culture, the Amish see little television and so are not so exposed to the reported five thousand daily commercial seducements and advertising that keep others agitated and wanting. They are not good prey for those who would take them.

The Amish also have a close and competent relationship with land, animals, and tools. They spend much of their time outdoors and so do not suffer from what Richard Louv calls "nature deficit disorder" and the boredom that thrives in a mostly indoor culture centered on the ephemeral and virtual. Their sanity, in other words, is grounded in a particular place and in the practical necessities of living, not so much in getting and spending. Psychologists, therapists, personal trainers, and the tribe of improvers who thrive in New York or Santa Monica do not prosper in Amish country.[35]

Finally, the Amish have controlled a problem largely ignored in the wider culture: they are not addicted to speed and constant movement and so are mostly immune to the great American motion sickness of high velocity, and its corollary, carelessness. They provide, accordingly, little incentive to the sellers of fossil energy, or the highway sprawlers and uglifiers. It is worth noting that no car or pickup truck has ever been destroyed by a horse-drawn Amish buggy, but sadly, the opposite is a common occurrence where trucks, cars, and buggies share the same roads. When faced with the choice between living extensively, expensively, and fast, or intensively, affordably, and deliberately, the Amish long ago chose the latter. The result is a people rooted in place who are neighborly, mutually supportive, and know how to make do with what is at hand.[36]

The Amish, imperfect as they are in some respects, are the best example we have of a sustainable and resilient economy and culture in the United States. But it is only one of many examples of cultural arrangements that have produced durable and prosperous economies without elaborate theoretical paraphernalia of the dog-chasing-its-tail kind. In his classic book *The Wheelwright's Shop*, George Sturt describes one of the last of the small firms that made wagons in rural Britain. His was a provincial business that catered to local farmers, so, he writes, "we got curiously intimate with the peculiar needs of the neighborhood . . . the dimensions we chose, the curves we followed were imposed upon us by the nature of the soil in this or that farm, the gradient of this or that hill, the temper of this or that customer or his choice perhaps in horseflesh." Without a supportive community, Sturt admits that he would have "soon been bankrupt . . . if the public temper then had been like it is now—grasping, hustling, competitive." But his clientele were reluctant "to take advantage of my ignorance" and he refused to lower the quality of the wagons he made. Sturt was part of a culture participating in "the age-long effort of Englishmen to fit themselves close and ever closer into England." It is a culture that has retained long memories, celebrated its place, and created "special products and skills, its peculiarities of cultivation, its delicacies and local dishes." Archeologist Jacquetta Hawkes writes, "I cannot resist the conclusion that the relationship [between people and land] reached its greatest intimacy, its most sensitive pitch about two hundred years ago. By the middle of the eighteenth century men had triumphed, the land was theirs, but had not yet been subjected and outraged." The common reaction is to dismiss such accounts as sentimental blindness, but as Hawkes writes, "it would be sentimental blindness of another kind to ignore the significance of its achievement—the unfaltering fitness and beauty of everything men made from the land they had inherited."[37]

I neither expect nor recommend that anyone having read the preceding paragraphs will rush out and join an Amish community or open a wagon-making business. I intend only to make the point once made by that great economist Kenneth Boulding, that "whatever is, is possible." Amish economies and many others similar to that described by Sturt exist or once did not so long ago. They exist, therefore, as possible examples of community-scale frugality, resilience, stability, velocity, and practical competence—all accomplished within a coherent culture and without much help from professional economists, experts, and counselors. In each case and in varying ways, competent people earned their livelihood at a human scale, one in which commerce can be shaped and comprehended as a harmonious part of the larger whole.

The problem, then, is not that we do not know about better alternatives, but that we don't yet feel them as real possibilities, and we don't feel them as anything more than idle curiosities because we still live in thrall to speed, accumulation, and convenience and believe that with still more clever technology we can get away with it—even as the biophysical and moral underpinnings of the old economy are disintegrating before our eyes. It is still easier, as someone recently remarked, for us to envision the end of the world than the end of capitalism. But the making of a better economy, capitalist or otherwise, hinges not only on our ability to see what is untried yet possible but also, and more importantly, on our ability to recollect older and sometimes saner ways of doing things that were not so much improved on as they were elbowed out of existence. Although talk of anything before, say, 1995 or any form of communication that works slightly slower than the speed of light is to risk being thought irretrievably daffy, it can be clarifying to note, as the Welsh anthropologist Alwyn Rees once observed, "When you have reached the edge of an abyss, the only thing that makes sense is to step back."[38]

At this late hour, standing on the edge of an abyss, is it possible to step back and still build a resilient, fair, prosperous, and durable economy? The only useful answer is a contingent "yes." Contingent because the laws, regulations, tax system, politics, media, advertisers, habits of the herd, presumptions of doctrine, failures of imagination, and the sheer power of money madness stand athwart better possibilities and urgent matters of human survival. Contingent, also, because we Americans, believing ourselves to be God's anointed, have not yet decided whether we wish one day to be known as history's most zealous and prodigious consumers and the Earth's most earnest and ingenious wreckers or whether we want to be celebrated for our art, literature, music, compassion, good humor, good schools, fairness, wise governance, great cities, the loveliness of our countryside, the health of our children, resilient prosperity, foresight, the life in our means of livelihood, and our fidelity to the cause of human dignity. Whichever path we choose, our economy, for better or for worse, will be the clearest portrait of who we are as a people and what we might—for better or worse—become. On such matters we remain divided.[39]

Our cultural default is to be "optimistic" and fervently believe that "the future is better than you think." Presumably it will be better partly because it requires no change in our aspirations and behavior, which is to say no deep improvement in us. The party, in other words, goes on, but with our sins absolved by "breakthrough" technologies. Such is optimism on the edge or maybe just another form of whistling past the graveyard at midnight. Pure optimists, as distinguished from the merely hopeful, have not a quiver of doubt about our capacity to get out of self-induced, technologically amplified, and increasingly complicated messes. They are happily situated in a wonderland of gee-whiz gadgetry that purifies, generates, grows, and makes money, which is, one may presume, the driver in the process. Despite the gravity of our situation, they are loath to "reign in our imaginations," which ironically mostly excludes creativity in the political, ethical, moral, or even economic realms. They are prone to celebrate, for example, the ingenuity of Coca-Cola's recent breakthroughs in water conservation even as that company prospers by depleting groundwater worldwide in order to sell more caffeinated sugared water to people who need fewer junk calories, much less sugar, but more hydration, nutrition, and local, not corporate, control of their lands, water, and lives. True to form, such optimists categorically dismiss dissenters as "Luddites" who might cause us to backtrack to the Amish level and thereby miss the "vast" benefits "afforded by all this technology." One suspects that the actual history of the Luddite movement remains unread along with the companion volumes on Fukishima, Bhopal, endocrine disrupters, nuclear weapons, gulags, and Auschwitz, as well as those yet to be written about the cozy and sinister relationship among the National Security Agency, the CIA, and the giant information technology companies. These pure optimists long for "breakthroughs," and with what must be either feigned innocence or world class naiveté, assume that if and when these occur they will have no hidden costs or impose no unforeseen collateral damages. They are technological fundamentalists whose manner of thinking, like that of other fundamentalists, is to offer one solution—more of the same—to all problems including those that are first and foremost deeply human, moral, and political. They do not deal in dilemmas that are not solvable.[40]

Sobriety on such portentous issues is more difficult in a culture where things are new and still shiny, and where few physical ruins and festering psychological wounds can remind the observant of human fallibility, stupidity, and, yes, evil. White America is still too young and too self-assured in the manner of a brash and callow adolescent to have been sufficiently tempered by the memory of screwups, malice, and the thousands of ways that things fall apart or that counterintuitive outcomes ruin the best-laid plans.

If real improvements—economic and other—do come, and I think they will, they must begin without the hallucinations and magical thinking that wishes away our history and wilfully overlooks the limits imposed by carrying capacity, complexity, and our own ignorance. If and when improvements happen, I believe they will start at the periphery of power and wealth, in the places and situations that are too small to be noticed or too insignificant to attract organized enmity: rather like the mouse-sized mammals that once scurried between the feet of dinosaurs, these initial experiments will scarcely draw public attention. In fact, the transition began decades ago in such places: neighborhoods and communities both rural and urban, here and elsewhere. It continues to gather force. But there is no overall strategy across a spectrum that includes people engaged in sustainable agriculture, slow foods, slow money, urban farmers, green builders, wind farmers, solar installers, bikers, inner-city businesses, environmental educators, pioneers and bioneers of all kinds, public servants, communicators, organizers, and those trying to level the economic playing field. In due course, the tide of change from the periphery might transform the larger culture and someday, perhaps, might cause tectonic political change as well.[41]

I leave it to others better equipped and wiser than I to describe the critical macro changes that must eventually be made at the state and national level in areas such as ownership, taxation, investment, finance, and public policy. But amid promising shoots and hints of change we will need a compass to clarify our central convictions and to remind us of what really matters. I know of none better than that once proposed by John Ruskin: "The great, palpable, inevitable fact—the rule and root of all economy—[is] that what one person has, another cannot have; and that every atom of substance, of whatever kind, used or consumed, is so much human life spent . . . for every piece of wise work done, so much life is granted; for every piece of foolish work, nothing; for every piece of wicked work, so much death." Accordingly a good economy would:

- exist within the limits of human scale;
- give priority to sufficiency over efficiency;
- designate times when nothing is bought or sold except to save lives;
- help people grow, not render them hooked, dependent, and gullible;
- include debt forgiveness, the fifty-year Jubilee, as a regular practice;
- provide good work that ennobles and dignifies;
- tax advertisements and ban those aimed to exploit children;
- prohibit all profits from the making and selling of weapons;
- distribute costs, risks, and benefits fairly;
- require that prices include the full life-cycle costs of the product or service;
- help us transition from a having culture to a being culture; and
- protect our common heritage of climate, wildlife, lands, and waters.[42]

These changes will not and cannot begin in the marketplace alone or in debates about how best to further self-interest at any scale. Instead, these conditions necessary to the survival of civilization will require a larger scaffolding of laws, regulations, and understandings that can mesh biophysical realities, intergenerational and interspecies morality, cultures, histories, and public engagement—something Herman Daly once described as macro control with micro variability. Only in this way will we ever hope "to make [human] madness as little harmful as possible."[43]

> But what is government itself but the greatest of all reflections
> on human nature? If men were angels, no government
> would be necessary. If angels were to govern men, neither
> external nor internal controls on government would be necessary.
> In framing a government which is to be administered by men
> over men, the great difficulty lies in this: you must first enable the
> government to control the governed; and in the next place,
> oblige it to control itself.
>
> —James Madison

> It is quite possible that by the year 2100 human life will
> have become extinct or will be confined to a few residential
> areas that have escaped the devastating effects of nuclear holocaust
> or global warming.
>
> —Brian Barry

The connection between the burning of fossil fuels and climate change was made in 1897. A full 118 years later, the world community finally bestirred itself at the COP-21 gathering in Paris to begin the long transition to a low-carbon future and hopefully cap off the worst possible effects of climate change. That agreement gives national leaders, legislators, financiers, corporate executives, and leaders at all levels standing to initiate long overdue changes in policy and finance. The Paris Agreement of 2015 may turn out to be a historic turning point. If so, it is only the beginning.

Meanwhile, the United States, which has just 4 percent of the world population and yet is responsible for an estimated 28 percent of CO_2 emissions historically, still does not have a binding national climate policy. Despite a flood of increasingly authoritative and urgent warnings about the dire consequences of inaction, it does not have a policy to tax carbon or one to cap emissions. This is not merely procrastination, an innocent sounding word, but in Gus Speth's words, "probably the greatest dereliction of civic responsibility in the history of the Republic." The Paris Agreement notwithstanding, the world is in the early decades of a protracted crisis—a witch's brew of worsening and unsolved ecological and social problems at different geographic and temporal scales all converging into a giant mashup. It is a collision of two "nonlinear" systems—the biosphere and biogeo-chemical cycles on one side, and human institutions and organizations on the other.[1]

Long before the climate crisis was "the greatest market failure the world has ever seen," as Nicholas Stern says, it was a massive political and governmental failure. The knowledge that carbon emissions might sooner or later threaten the survival of civilization was known or suspected decades ago, but the federal government has only recently bestirred itself to act. The reasons for governmental lethargy are many. Climate change is the "perfect problem": scientifically complex, politically divisive, economically costly, morally contentious, and ever so easy to deny or to defer to others at some later time.

But there is another reason that climate change is being ignored. For half a century, a concerted war has been waged against government in the Western democracies, particularly in the United States and the United Kingdom. Its origins can be traced back to the more virulent strands of classic liberalism that were once arrayed against the entrenched power of royalty. Its present form was given voice by Ronald Reagan, who reoriented the Republican Party and U.S. politics around the idea that "government is the problem," and by Margaret Thatcher in Britain, who ruled in the firm conviction that there was no such thing as society, only atomized self-interests. Other forces and factions joined in an odd alliance of ideologists, corporations like ExxonMobil, media crackpots, and economists like Milton Friedman, who called themselves "conservatives," which, all things considered, is an odd choice of words.[2]

Other factors have contributed to the hollowing out of Western-style governments. In the United States in particular, wars and excessive military spending have contributed greatly to deficits, the impoverishment of the public sector, and the declining credibility of public institutions. The rise of multinational corporations, too, has created rival sources of authority and power. Electoral corruption, gerrymandering, and right-wing media have fanned the flames of public hostility toward governments, politics, full democracy, and even the idea of a public good. Defunded and harassed, morale and effectiveness declined throughout many federal agencies, thereby reinforcing the belief that governments are always incompetent compared to the market, which was, in fact, created by public choices. The Internet has helped as well to balkanize the public into ideological tribes that are seemingly incapable of a civil dialogue.

But the war against government is not what it is purported to be. In fact, it is not a war against excessive government at all, but rather a well-funded campaign to reduce only those parts of government dedicated to improving public health, transportation, education, quality of the environment, and our common infrastructure.[3] The warriors of the right, once fiscal conservatives, without a twitch of doubt will cut taxes on corporations and the wealthy and lavishly subsidize fossil-fuel industries while eagerly borrowing money for wars, weapons, and domestic surveillance, all of which will add to the very national debt they otherwise deplore.

As a result of this assault on the very idea of government, our capacity to solve public problems has diminished sharply and the power of private-sector banks, financial institutions, and corporations has risen. Not surprisingly the countervailing and regulatory power of democratic governments has been eroded, and with it much of the effectiveness of public institutions to foresee, plan, and act—that is, to govern.

The erosion of the governing capacity in the United States is happening at a very bad time. We are still the most militarized and wealthy nation-state on Earth, but our influence for good is in decline due in large measure to the catastrophe that began with the invasion of Iraq in 2003. Aside from the death toll on all sides and the cost of perhaps up to $3 trillion, the result was further destabilization of an already unstable region, which in turn led to the rise of ISIS and the spread of terrorist groups outside the region. The global system will be increasingly stressed by rapid climate change, continuing population growth and persistent poverty, and gross inequalities in wealth. Such problems are further exacerbated by the fact that some states are armed with nuclear weapons, some are clinging to ancient religious and ethnic hatreds, and still others are holding fast to their economic and political advantages. The combination threatens to undo order of any kind.[4] On the not-too-distant horizon are other and totally unprecedented challenges to governance posed by the rise of artificial intelligence systems that have "the potential to become the ultimate controllers of everything" and so may require "a radical reinvention of our political systems."[5]

Climate change has already caused severe damage in many parts of the world, but looking to the future, worse is coming. Warmer and more acidic oceans will be less capable of supporting humankind. Larger storms, rising seas, higher temperatures, prolonged drought, and disassembling ecologies are already wreaking havoc on agriculture, water systems, public health, and the urban infrastructure that provides transportation, electricity, and emergency services. Climate destabilization will grow worse for a long time to come. Even if we manage to stabilize carbon-dioxide levels by midcentury, the effects of the elevated emissions will last for centuries, perhaps millennia, and no society, economy, or political system will escape the consequences.[6]

We should brace, then, for a "perfect storm" of catastrophes:

1. Because of the combustion of fossil fuels and changes in land use, the Earth will undergo a substantial and rapid warming of not less than, and probably more than, two degrees Celsius—a change that will last for hundreds and possibly thousands of years.
2. Rising temperatures will progressively destabilize the global system, causing massive droughts and temperature extremes,

more frequent storms, rising sea levels, loss of a large fraction of species, and unpredictable ecological shifts.[7]

3. Climate destabilization will likely result in scarcities of food, energy, and resources, which will undermine political, social, and economic stability and amplify the effects of terrorism and conflicts between and within nations, failed states, and regions.

4. The point at which Earth's climate might cross into a runaway state of decline is unknown, but there is good scientific reason to believe that the threshold of irreversible catastrophe would be crossed at an atmospheric carbon dioxide concentration of between 400 and 500 parts per million. It is presently more than 402 ppm and rising at over two parts per million a year. The nightmare scenario involves a warming level high enough to thaw tundra and warm oceans enough to release vast amounts of methane, a more potent but shorter-lived greenhouse gas than CO_2.[8]

5. In a tightly coupled global system, the likelihood of unpredictable, significant ("black swan") events with irreversible, long-term global consequences will continue to rise.[9]

6. Because oceans still act as a thermal anchor in the Earth system, the recent climate-change-driven weather anomalies are the result of CO_2 emissions from about thirty years ago. The buffering effect will decline as oceans continue to warm and become more acidic.

There are presently no technologies that can remove approximately nine billion tons a year of carbon from the atmosphere and that are themselves carbon-neutral, affordable, and deployable at the scale and speed necessary to avoid serious global disruption. Geoengineering the atmosphere poses unknowns far beyond the scope of our current knowledge and forecasting abilities, and at best would only defer problems to a later time. Action to head off the worst of what could occur is difficult because of the complexity of nonlinear systems with large delays between cause and effect, and because of the political and economic power of fossil-fuel industries to prevent timely corrective action that would jeopardize their considerable profits.[10] Unfortunately, the effects of our procrastination will fall with increasing weight on coming generations, constituting the largest moral lapse in history.[11] For climate destabilization is not just an issue of technology and policy, but also a symptom of deeper problems rooted in our paradigms, philosophies, popular delusions, and the way our brains have evolved to handle discrepant and threatening information.[12]

The perfect *political* storm looming ahead will in turn be caused by the convergence of unavoidably worsening climate change; expanding ecological disorder (including deforestation, soil loss, water shortages, species loss, and the acidification of oceans); population growth; an increasingly unfair distribution of the costs, risks, and benefits of economic growth; as well as national and ethnic tensions— all compounded by the problems of political incapacity and human fallibility. Specifically, governments in the decades ahead will have to address the consequences and costs of:

- more prolonged and severe droughts as rivers, wells, and aquifers dry up;
- crop failures and more frequent famines;
- emergency relief in the wake of climate-change-driven storms, tornadoes, hurricanes, tsunamis, floods, and dust storms;
- the relocation of millions of people from regions affected by the rise in sea levels, flooding, or prolonged droughts;[13]
- the management of climate refugees, estimated to number 250 million or more by 2050;
- growing domestic and international conflicts over water and food, compounded by ancient ethnic conflicts and rendered more deadly by the diffusion of biological and nuclear weapons;
- power outages due to lack of cooling water and/or fuels;
- the increasing likelihood that economic growth will falter, particularly where it is needed most;
- maintaining public order under conditions of growing unrest;

- managing disputes such as whether to grow fuels for the wealthy or food for everyone else;
- building more resilient food, energy, and water systems;
- dealing with "stranded assets," notably fossil fuel resources that cannot be burned without the risk of global catastrophe and without losing many of the costs sunk into the infrastructure for extracting, processing, transporting, distributing, and burning oil, gas, and coal; and
- creation of a fair, sustainable, and resilient social and economic order.

Avoiding the worst that could happen will require sharp reductions of CO_2 emissions, trending toward zero emissions by midcentury. We are possibly close to the threshold beyond which climate change will be uncontrollable no matter what we do. To have any chance of dodging that bullet, we will have to quickly sequester the remaining reserves of fossil fuels that cannot be safely burned. Our choices are roughly to:

1. Confiscate fossil fuels from their present owners.
2. Compensate their current owners, rather like the British did when they ended slavery in the Caribbean in the nineteenth century.
3. Rapidly deploy alternative technologies and thereby render fossil fuels uncompetitive.
4. Geoengineer the atmosphere in order to lower temperatures and buy time to think of something better to do.
5. Some combination of all of these strategies.[14]

The particularities and perplexities of policy aside, if civilization is to last, we must permanently remove reserves of coal, oil, tar sands, and natural gas from the asset side of the economic ledger in ways that will not capsize the global economy.

No government has yet shown the foresight, will, creativity, or the capacity to deal with systemic problems at the scale, complexity, or duration necessary given how long carbon dioxide remains in the atmosphere. No government is presently constituted or competent enough to make the "tragic choices" or unavoidable triage decisions required in ways that are humane and rational. To the contrary, "conservatives" in the United States and elsewhere, devoted as they are to the libertarian philosophy and greatly indebted as they are to the oil, gas, and coal companies, have been busy dismantling governments' capacity to solve large-scale problems, many of which have been caused or worsened by their benefactors. And no government has yet demonstrated the capacity to rethink its own mission at the intersection of climate instability and the conventional economic wisdom embedded in doctrines of "neo-liberalism," the "Washington consensus," and the prevailing devotion to economic growth. The same is true in the realm of international governance. In the words of historian Mark Mazower: "The real world challenges mount around us in the shape of climate change, financial instability . . . [but there is] no single agency able to coordinate the response to global warming."[15]

James Madison's contributions to the *Federalist Papers* and other writings are among the most brilliant reflections ever written on the art and science of governance. He was not optimistic about the long-term prospects for democracy in part because of the problem of controlling what he called "factions" organized around competing interests. His proposed solutions included diluting factions in a larger, more diverse nation; structuring government so as to oppose one faction against another; and dividing power among the three branches of the federal level and between the national government and that of the various states. The result was an adversarial system in which rapid, large changes would be difficult to make and no faction or branch of government could easily dominate the others. But Madison's prescriptions had a kind of self-fulfilling dynamic that would promote the very problems he aimed to solve. In political scientist Steven Kelman's words: "Design your institutions to assume self-interest . . . and you may get more self-interest. And the more self-interest you get, the more draconian the institutions must become to prevent the generation of bad policies."[16]

Humans are motivated by many things, and in different political cultures at different times self-interest can be narrowly or broadly conceived. Political institutions and policies accordingly ought to be designed, whenever possible, to nudge behavior toward a common good without overt coercion.[17]

Were James Madison to return now, he would be astonished that 226 years later the government he had helped to design for an agrarian age is still functioning—more or less—in an industrial-technology-driven world of more than seven billion people. He would also be dismayed, I believe, by the backlog of unsolved problems accumulating in all societies, most of which can be attributed to governmental and political failures of one kind or another. But he would not be surprised by the present disenchantment with government, or the hyper-factionalism that has prevented us from solving our most pressing problems. Facing prospects miniscule in comparison to ours, Madison was nonetheless "driven by a nightmare," in historian Richard Matthews's words, and believed that the experiment in governance he would help to launch would be doomed to failure, perhaps within a century.[18]

Madison's great achievement as a political thinker was to astutely analyze the fatal flaws in the Articles of Confederation, clarify the dire consequences looming ahead for the Confederation, and articulate cogent and novel principles necessary for a much stronger central government. But for all of its brilliance, the Constitution written in 1787 had serious shortcomings. To secure its passage, the authors avoided the issues of slavery, America's most original and egregious sin. They ignored and thereby excluded women and native peoples. They privileged property owners and the wealthy. They chose methods of election to the presidency and senate that were less than democratic. And they gave too much power to small (population) states and underrepresented urban states. In time some of the flaws of the Constitution were remedied, though others still fester.[19]

Moreover, Madison could not have foreseen that the problem of faction would one day grow into a particularly pernicious and seductive form of tyranny: the limited liability corporation. In time, the corporation would become a kind of Trojan horse corrupting democracy, buying undeserved influence, demanding massive subsidies, deflecting economic development, undermining the free press, distorting free elections, creating a resource-intensive consumer society, and threatening human survival itself.

Even worse, nearly a century after the *Santa Clara County* precedent, the U.S. Supreme Court ruled in *Buckley v. Valeo* (1976) that the expenditure of money in political campaigns is a form of free speech subject to the protections of the First Amendment. And in *Citizens United v. FEC* (2010) the Court ruled that there are no limits on campaign expenditures and no legal requirement for disclosure. Consequently, the flood gates have been opened wide for large, secret campaign contributions by both individuals and corporations—contributions that undoubtedly have influenced the political system in profound and undemocratic ways.

Madison and the founders could not have foreseen the scale or corrupting effects of large corporations. But a future unimaginable to them has become our reality and corporations are now the dominant features on the political and economic landscape. The question of whether and how they fit a democracy remains unanswered, however. One of the shrewdest students of politics and economics, Yale political scientist Charles Lindblom, concluded his book *Politics and Markets* with the observation that "the large private corporation fits oddly into democratic theory and vision. *Indeed, it does not fit.*" Until democratized internally and rechartered to be made publicly accountable and to serve publicly useful purposes, corporations will remain autocratic, powerful, undemocratic fiefdoms within what is ostensibly a democratic republic.[20]

The passage of time has revealed other serious Constitutional flaws:

> The U.S. Constitution fragments power and so renders environmental lawmaking difficult, or as legal scholar Richard Lazarus explains, it provides no "clear, unambiguous textual foundation for federal environmental protection law." Instead, it privileges "decentralized, fragmented, and incremental

lawmaking . . . which makes it difficult to address issues in a comprehensive, holistic fashion." Congressional committee jurisdictions established by the Constitution further fragment responsibility, perspective, and legislative results.[21]

The Constitution gives too much power to private rights as opposed to public goods. It does not mention the environment or the need to protect soils, air, water, wildlife, and climate and so offers no unambiguous basis for environmental protection.

The commerce clause, the basis for major environmental statutes, is a cumbersome and dubious legal basis for environmental protection. The result is that "our lawmaking institutions are particularly inapt for the task of considering problems and crafting legal solutions of the spatial and temporal dimensions necessary for environmental law."[22]

They are inapt as well for protecting the lives, health, and environment of our descendants. Posterity is mentioned only once in the Preamble to the Constitution, but not thereafter. The omission is understandable in their time, but in ours it is an egregious wrong. Theirs was an agrarian world in which the fastest manmade thing on Earth was a sailing ship in a strong wind. The most lethal device was a cannon that could hurl a twenty-pound solid iron ball less than half a mile. They could have had no inkling that far into the future one generation could deprive all others of life, liberty, and property without due process or even without good cause. And so, in Thomas Berry's words: "It is already determined that our children and grandchildren will live amid the ruined infrastructures of the industrial world and amid the ruins of the natural world itself." The U.S. Constitution gives them no protection whatsoever.[23]

In his 1974 book *An Inquiry into the Human Prospect*, economist Robert Heilbroner wrote: "I not only predict but I prescribe a centralization of power as the only means by which our threatened and dangerous civilization will make way for its successor." Heilbroner's description of the human prospect included not only global warming but also the wide range of other threats to industrial civilization, including the possibility that finally we would not care enough to do the things necessary to protect posterity. The extent to which power would have to be centralized, however, would depend greatly on the capacity of populations accustomed to affluence and consumption to muster "willing self-discipline." But he did "not find much evidence in history—especially in the history of nations organized under the materialistic and individualistic promptings of an industrial civilization—to encourage expectations of an easy subordination of the private interest to the public weal." In the end Heilbroner, like many others, found little cause for optimism. His conclusions are broadly similar to those of others including British sociologist Anthony Giddens who, somewhat less apocalyptically, proposes "a return to greater state interventionism" but with the state acting as a catalyst, facilitator, and enforcer of guarantees. Giddens believes that the climate crisis will motivate governments to create new partnerships with corporations and civil society, which would be more of the same only bigger and better.[24] Robert Rothkopf likewise argues that the role of the state must evolve toward larger and more innovative governments and "stronger international institutions [as] the only possible way to preserve national interests." In this regard the U.S. government lags behind those of virtually every other capitalist society.[25]

The history of highly centralized governments and corporations, however, is not encouraging, especially relative to the conditions of the long emergency. Governments have often been effective at waging war. Sometimes they are effective in solving economic problems. But the historical record indicates that they are mostly cumbersome, slow, ineffective, and overly bureaucratic. Might there be faster, more agile, and less awkward ways to stabilize the climate and achieve common purposes without authoritarian governments, the inevitable compromises and irrational messiness of politics, or reliance on personal virtue and sacrifice? And if so, can these be made to work over the long time spans necessary to restabilize climate conditions? Broadly, I see three possibilities.[26]

The first comes from those champions of markets and advanced technology who propose to solve the climate crisis by harnessing the power of markets and technological innovation and avoiding what they regard as the quagmire of government. Rational corporate responses to markets and prices, they believe, can achieve many of the same environmental goals faster and at lower costs and without hair-shirt sacrifice, moral posturing, or slow, clumsy, overbearing bureaucracies. This potential solution is predicated on a combination of informed self-interest and the revolution in energy efficiency and renewable energy that has made renewables cheaper, faster, less risky, and more profitable than reliance on even highly subsidized fossil fuels. In their tour de force *Reinventing Fire*, Amory Lovins and his co-authors, for example, ask whether "the United States could realistically stop using oil and coal by 2050? And could such a vast transition toward efficient use and renewable energy be led by business for their pecuniary advantage?" The answer, they say, is "yes" and the reasoning and data they marshal are formidable.[27]

But a reliance on markets to resolve the issues driving the long emergency raises many questions. Why would any corporation, particularly one in a highly subsidized extractive industry, agree to change as long as profits are high and it can pass on the costs of climate change to others now or later? In this regard the history of ExxonMobil is instructive. Would corporations act for the public good if they were not forced to do so by the police power of the state? Would oil and coal companies demand compensation for leaving their stranded assets unburned, even when necessary to avoid runaway climate change? And would greener, energy efficient solar-powered corporations continue to use their financial power to manipulate public opinion here and elsewhere, undermine regulations, and oppose other changes necessary to the equitable sharing of costs, risks, and benefits in the long emergency? Does the same logic apply throughout the global economy? However one answers such questions, the fact is that when climate-change-driven disasters hit, as they will, the victims are unlikely to call 1–800-Wal-Mart. They will instead call 911 or its equivalent and desperately hope that someone answers—preferably someone who is competent, organized, well-equipped, and working in a well-administered federal agency.[28]

Further, what kind of society do we intend to build? It is possible to imagine a corporate-dominated, hyperefficient, solar-powered, sustainable, and fascist society. Some believe that is where we are heading—an Ayn Randian hell organized solely by market transactions. The result, as Karl Polanyi once warned, would demolish society. But to repeat the point, there are some things that should never be sold—sometimes because the selling undermines basic human rights, sometimes because it violates the law and procedural requirements for openness and fairness, sometimes because the selling would have a coarsening effect on society, sometimes because the sale would steal from the poor and vulnerable including future generations, sometimes because the thing to be sold is part of the common heritage of humankind and so can have no rightful owner, and sometimes because some things, including government itself, should not be for sale, ever, under any circumstances.[29]

A second alternative to authoritarianism is described by Paul Hawken as "blessed unrest"—a global upwelling of groups, foundations, organizations, and networks working in various ways to promote sustainable economies, solar energy, justice, transparency, and further community mobilization as a way of building a new global order from the bottom up. Many of the thousands of groups that Hawken describes are linked in what Steve Waddell calls "global action networks," which are organized around specific issues to provide "communication platforms for sub-groups to organize in ever more specialized geographic and sub-issue networks." The earliest examples of such networks include the International Red Cross and the International Labour Organization. More recently, action networks have formed around issues of managing common-property resources, global financing for local projects, water, climate, political campaigns, and access to information. With the internet, global action networks are becoming ubiquitous. They are fast, agile, participatory, and focused. Compared to other citizen-initiated efforts, their costs for overhead and administration are low. But like other grassroots activities, they have no power to enforce, tax, or defend, which are prerogatives of the state, and because they depend heavily on volunteers and underpaid employees, often they lack staying power. In Mark Mazower's words, "Many are too opaque and unrepresentative to any collective body." Regarding the efforts of "philanthrocapitalists," Mazower believes that by "applying business methods to social problems, they exaggerate what technology can do, ignore the complexities of social and institutional constraints, often waste sums that would have been better spent more carefully and wreak havoc with the existing fabric of society in places they know very little about." Moreover, they are no less immune to corruption and arrogance than governments, but they operate with little or no accountability or transparency.[30]

This brings me back to Heilbroner. Is it possible that he was wrong not in his assessment of the kinds of things that we will have to do to improve the human prospect, but in his assessment of democracy? The authors of the U.S. Constitution grounded ultimate power in "we the people," while mostly denying them any such power or much access to it. Further, the Constitution is founded on two contradictory principles. On one hand it fragmented and carefully reined in the powers of the national government. On the other, it gave states powers and the right to exercise the "kind of coercion forbidden to the central government." That contradiction, in historian Gary Gerstle's view, "may portend the nation's decline" due to Congressional paralysis, the role of corporate and unaccountable private money used to influence states' policies, and a Supreme Court decision that limits the power of the federal government to implement policy in the states. Further hamstringing the government's response, "unremitting hostility to the exercise of public power at the federal level . . . has all but paralyzed its ability to address problems confronting the country in the twenty-first century."[31]

The fact is that our framing document was not and still is not very democratic. As Harold Myerson describes it, "the problem isn't that we're too democratic. It's that we're not democratic enough." In My-erson's view the problem is that long ago we ceded whatever power we had to corporations, oil companies, bureaucrats, and managers of one kind or another in return for bovine acquiescence. In a growing economy, such subservience seemed to work pretty well for most people most of the time. In a slow growing or stagnant economy, however, the bargain unravels. Are we to assume that those who prospered and reigned on the upslope will serve the public interest on an economic plateau or the downslope?[32]

Columbia University professor and *New York Times* columnist Thomas Edsall thinks it unlikely: in his view, "the American conservative movement is taking preemptive action appropriate to a slow but steady collapse." With the scent of decline in the air, "the contemporary economic climate has dealt a blow to the instinct of generosity essential to the left and has strengthened the instincts of greed, callousness, and self-preservation." The reasons are deeply embedded in human nature and traits of greed, fear, and shortsightedness that are now amplified by climate destabilization.[33]

So what is to be done? Canadian writer Naomi Klein describes the agenda ahead:

> Responding to climate change requires that we break every rule in the free-market playbook and that we do so with great urgency. We will need to rebuild the public sphere, reverse privatizations, relocalize large parts of economies, scale back overconsumption, bring back long-term planning, heavily regulate and tax corporations, maybe even nationalize some of them, cut military spending and recognize our debts to the global South. Of course, none of this has a hope in hell of happening unless it is accompanied by a massive, broad-based effort to radically reduce the influence that corporations have over the political process. That means, at a minimum, publicly funded elections and stripping corporations of their status as "people" under the law.[34]

"Climate change," she continues, "detonates the ideological scaffolding on which contemporary conservatism rests. There is simply no way to square a belief system that vilifies collective action and venerates total market freedom with a problem that demands collective action on an unprecedented scale and a dramatic reining in of the market forces that created and are deepening the crisis."

It is virtually impossible to imagine such things happening democratically without a significant deepening and transformation in the practice of democracy itself. Political theorist Benjamin Barber calls the transformed version "strong democracy," by which he means a "self-governing community of citizens who are united less by homogeneous interests than by civic education and who are made capable of common purpose and mutual action by virtue of their civic attitudes and participatory institutions rather than their altruism or their good nature." This sort of participatory democracy of engaged, thoughtful citizens is along the lines once proposed by Thomas Jefferson and later by John Dewey. The primary obstacle to a strong democracy, Barber notes, is the lack of a "nationwide system of local civic participation." To fill that void he proposes, among other things, a national system of neighborhood assemblies.[35]

There are many other proposals to strengthen democratic institutions and the habits of citizenship. Amy Gutmann and Dennis Thompson, for example, propose a "deliberative democracy" consisting of publicly funded institutions in which "free and equal citizens (and their representatives) justify decisions in a process in which they give one another reasons that are mutually acceptable and generally accessible, with the aim of reaching conclusions that are binding in the present on all citizens." They propose to get Americans talking about large issues in public settings, including corporations and schools, in ways reminiscent of classical Greek democracy and more recently in proposals by sociologist Jürgen Habermas. They intend to increase the legitimacy of critical choices, improve public knowledge, and increase civil discourse. A great deal depends, as they concede, on "whether proponents can create and maintain practices and institutions that enable deliberation to work well." They propose to extend deliberative practices throughout civil society but consider schools, colleges, and universities to be the most important. Political scientists Bruce Ackerman and James Fishkin, similarly, propose to establish a new national holiday called "Deliberation Day," in which citizens would meet and engage in structured dialogues about issues and candidates. They believe that "ordinary citizens are willing and able to take on the challenge of civic deliberation during ordinary times" in the kind of properly structured setting that "facilitates genuine learning about the choices confronting the political community."[36]

Legal scholar Sanford Levinson goes further, recommending a national convention to remedy the structural flaws that make the U.S. Constitution less democratic than any of the fifty state constitutions. If not a call to revolution, he calls for active citizenship to reform the parts of the Constitution that render the future more difficult than it otherwise might be. Better leadership, more engaged citizens, a more robust political culture, changes to the media, and campaign finance reform—good ideas all—but ineffective and enacted in vain unless we overcome serious flaws in the structure of our Constitution. It is a flawed document that was designed to make change difficult and to protect entrenched interests.[37]

Despite the calls for reform, the question is whether people here or in other democracies are capable of the kind of citizenship necessary to play a constructive role in the long emergency. A possible answer can be found in the troubled history of democracy itself and the complexities of political theory. Classical democracy in Athens, limited as it was, lasted for only two hundred years. "Democracy," as political philosopher John Plamenatz once wrote, is "the best form of government only when certain conditions hold." But many of those conditions, already weakened, may not hold in the long emergency ahead. The reasons for this are many.[38]

For one, modern democracies arose in what historian Walter Prescott Webb once described as the "great frontier," in which land, energy, and resources were abundant. American democracy, in particular, owes a great deal to the abundance of land and resources. We became a "people of plenty" in historian David Potter's words, accustomed to comfort and affluence. But the long emergency will require discipline and sacrifice, traits no longer as common as they once were. How democracy will fare in more straitened times is an open question. In short, no one knows whether democratic processes or governing structures can evolve to manage the challenges ahead. Political analyst Peter Burnell cautions that "democratization does not necessarily make it easier and can make it more difficult for countries to engage with climate mitigation."[39]

Even at their best, representative democracies are vulnerable to neglect, changing circumstances, corruption, and the frailties of human judgment. They tend to atrophy, becoming ineffective and corrupt agents of the powerful and wealthy. Democracies, in particular, are vulnerable to ideologically driven factions that refuse to play by the rules of compromise and fair play that are requisite to stable democracies. They are susceptible to all of the forces that corrode political intelligence. "Why is it," Erich Fromm once asked,

> that people cannot see the most obvious facts in personal and social affairs and, instead, cling to clichés which are endlessly repeated without ever being questioned? Intelligence, aside from the native faculty, is largely a function of independence, courage, and aliveness; stupidity is equally a result of submission, fear, and inner deadness.[40]

Democracies are also vulnerable to what conservative philosopher Richard Weaver once described as the "spoiled-child psychology": "a kind of irresponsibility of the mental process . . . because [people] do not have to think to survive . . . typical thinking of such people [exhibits] a sort of contempt for realities." There is good evidence that the spoiled-child behavior that Weaver noted in the 1940s has morphed into a full-blown riot of self-indulgence—psychologists Jean Twenge and Keith Campbell describe an "epidemic of narcissism" today, by which they mean "a giant transfer of time, attention, and resources from reality to fantasy . . . corrod[ing] interpersonal relationships." The upshot is "a switch from deep to shallow relationships, . . . destruction of social trust, and an increase in entitlement and selfishness." In societies saturated 24/7 with fantasy and entertainment, it should come as no surprise to the alert that we are on a "collision course with reality . . . [and] reality may be losing if it has not done so already."[41]

GOVERNANCE

We've come to an impasse. There is, in short, no good case to be made for smaller governments unless we are willing to reduce our expectations for security and prosperity in an increasingly insecure age and lower our expectations of the good society toward a libertarian, gun-toting, and bloody free-for-all. If we intend otherwise we will need representative governments to be active, creative, and effective agents—which can happen only with astute and constant care, public engagement, civic intelligence, tolerance, willingness to compromise, a bit of wisdom now and then, and public vigilance as both Madison and Jefferson originally intended. In democracies, governments are the instrument through which the public decides issues of justice, fairness, equality, environmental quality, public health, equal opportunity, war and peace, and public investment in things that markets do not and cannot provide. Properly led, managed, and funded, governments do things that we cannot otherwise do on our own—tasks that require collective solutions and foresight. But as Madison noted, there are requisites for capable and competent governments, including (1) a genuinely free press that serves a well-informed and engaged citizenry, (2) free, fair, and open elections, and (3) reliable ways to counterbalance various factions and interest groups. I add two others.

The first is to connect the practice of democracy with dramatic advances in ecological design and biomimicry that are transforming fields as diverse as agriculture, architecture, engineering, urban design, transportation, and manufacturing. In combination with the revolution in energy efficiency and renewable energy, it is now technically and financially possible to design cost-effective buildings, neighborhoods, and cities powered mostly or entirely by sunlight and to reduce costs while increasing conviviality, prosperity, resilience, and local control. It is becoming possible to eliminate virtually all waste and most industrial pollution by manufacturing methods that mimic natural systems. It is possible, then, that a combination of distributed, renewable energy; local agriculture; and ecologically grounded design could profitably transform the economy and significantly reduce the need for government regulation while appealing for different reasons to both conservatives and liberals. These are no longer distant possibilities, but emerging realities that could dramatically improve the design of buildings, food systems, transportation networks, manufacturing facilities, and urban areas to better harmonize with natural systems *and* improve the quality of life and the economy without having to create the scale of government that Heilbroner thought necessary but that no one wants. It is possible, in other words, to design practical and economically smart solutions that eliminate pollution, speed the transition to renewable energy, and reduce the need for government regulation and federal spending. The government's role in this scenario would be much like the one it has taken in the internet-based areas of the economy: that of an initiator, facilitator, and early investor in research and deployment, not the owner and manager.[42]

A second change comes straight through the centuries from Thomas Jefferson, who envisioned a rural society of citizen farmers who owned their land. The kernel of the idea is widespread ownership so that no one would be subject to arbitrary and unaccountable power. In our time it means spreading democracy through the economy and, most important, through those autocratic fiefdoms otherwise called corporations. Worker ownership is an old idea and it has worked well to the advantage of workers, management, and the firm. Why should the practice of democracy stop at the corporate or factory front door?

Outside of Hollywood films, stories don't always have happy endings. The human record, to the contrary, is a catalog of error, tragedy, injustice, and stupidity, punctuated by small and often paradoxical improvements, as well as the occasional triumph. As an undergraduate history major once famously observed, it's "one damn thing after another." And one of those damn things is periodic collapse. When they don't see large problems ahead or summon the wit to solve them in time, entire societies and civilizations collapse in various ways. Whatever the particulars, the downward spiral leading to collapse always includes a large dose of incompetence and irresponsibility by the elites, who are often engaged in wishful thinking, denial, and groupthink abetted by rules that reward individual selfishness, not group success. In other words, the collapse of societies and civilizations begins with failures of governance and politics that have to do with "who gets what, when and how."[43]

The long emergency is no exception. Its causes are first and foremost political, not technological or economic, but they point to something deeper: "a profound human crisis that cuts to the heart of our civilization." In short, any chance that we have of surviving the trials of climate destabilization in a nuclear armed world of perhaps up to eleven billion people will require that we immediately reckon with the thorny issues of governance, politics, and the larger purposes they serve with the same kind of devotion, persistence, wisdom, and boldness that we associate with, say, the founding of the United States, the end of apartheid in South Africa, or the creation of the European Union. Difficult? Absolutely. But tackling these problems will be a great deal easier than trying to maintain an unworkable status quo.[44]

The fact is that we're plundering the Earth, taking far more than it can give—and often for trivial purposes—then distributing the benefits unfairly. We are robbing future generations of their life, liberty, property, and right to live in a fecund and beautiful world. But it is possible for us to do otherwise, to live elegantly and fairly within the limits of nature. If that is what we intend we will need a broader, deeper, and more modest sense of who we are. Our laws and constitutions are written as if they pertained only to the current generation and only for humans, their affairs, and property. A more inclusive and accurate view would embrace all of creation and extend rights of sorts to species, rivers, landscapes, and trees as legal scholar Christopher Stone once proposed. In Thomas Berry's words: "We have established our human governance with little regard for the need to integrate it with the functional order of the planet itself," leaving "the Earth defenseless against the savage assaults by its human inhabitants and their corporate embodiment." In fact, from our bodies to our global civilization we are part of a worldwide federation of beings, systems, and forces beyond our understanding. We are kin to all that ever was and all that ever will be. Our bodies are a congress of many organisms, shot through with wildness. We are made of stuff that exploded out of stars, and are products of millions of years of evolution. Our tribes, societies, and nations live by the beneficence of natural systems. Recognizing and appreciating this deep interconnectedness should inspire and compel us to extend compassion, care, and protection beyond our small circle.[45]

seven
MIND

only connect

—E. M. Forster

Thinking means connecting things, and it stops if they cannot be connected.

—G. K. Chesterton

Gyre: a circular movement or turn; a revolution; a circle; a spiral; a vortex

—*Oxford English Dictionary*

Fifteen-hundred miles west of Seattle, in the middle of the North Pacific, a mass of plastic debris and chemical sludge is caught in ocean currents known as the North Pacific Gyre. It is estimated to be the size of the lower forty-eight states and to be suspended at a depth of a hundred to a thousand feet. No one knows for certain how large or how deep the garbage gyre is, only that it is massive and growing. Some of the most amazing things that humans have ever made float in it. They are made primarily of oil extracted from deep below the surface of the Earth, which is another remarkable story. The impact on marine organisms and sea life is poorly documented, but it is considered to be somewhere between disastrous and catastrophic. Some of the debris is ingested by birds and fish that mistake floating plastic doo-dads for food. Some of it breaks down into long-lived toxic compounds. Despite its size and ecological effects, however, the North Pacific Garbage Gyre is remote enough to be out of sight and out of mind.[1]

Another gyre of gases circulates around the Earth six miles above our heads, the result of our annual combustion of four cubic miles of primeval goo—ancient sunlight congealed in the form of coal, oil, natural gas, shale oil, and tar sands. The atmospheric residues, chiefly CO_2, are now over 402 parts per million—said to be the highest concentration of the past 800,000 years, but perhaps of the past several million years. The atmospheric CO_2 gyre is changing the thermal balance of Earth in an instant of geologic time and locking us into a future of extreme heat, drought, larger storms, rising sea levels, and changing ecologies that will increasingly imperil economies, public health, and social and political stability.

A third gyre of long-lived chemicals cycles through our bloodstream, and some are stored permanently in our fatty tissues. They are in our air, water, food, everyday products, and many children's toys. In the words of the President's Cancer Panel, babies are born "pre-polluted," poisoned by toxic substances that pass through their mothers' umbilical cords. A typical sample of chemicals in the average body would include two hundred or more that are suspected or known to cause cancer and cell mutations and to disrupt the endocrine system. It is possible that, singly or in combination, invasive chemicals also cause behavioral abnormalities. Since the Environmental Protection Agency studies the effects of chemicals one by one, we don't know much about the possible combined effects of the tens of thousands of chemicals to which we are exposed or the several hundred that we've ingested, absorbed, and inhaled.[2]

The three gyres have many things in common. They are vicious cycles or "wicked problems" that are complex, long-term, and nonlinear—a fancy way to say they are unpredictable with lots of unknowns. They involve virtually every discipline listed in a college catalog and much outside the conventional curriculum as well. But they are not so much problems that can be solved with enough money and effort as they are dilemmas that could not, and likely cannot, be solved. With foresight, however, each could have been avoided.[3]

The effects of each gyre will last for a long time. Toxic and radioactive trash will threaten human health and ecologies for centuries to come. The loss of biodiversity driven by climate change, pollution, and overdevelopment is permanent. Carbon dioxide in the atmosphere will affect the Earth's climate for thousands of years, requiring a level of public and private vigilance for which we have no historical precedents. Heavy metals and persistent organic chemicals last a lifetime in the human body, and some are passed on to our offspring.

The causes of each gyre were known a long time ago. It required no great prescience to see that our mountains of trash would someday return to haunt us. The adverse health effects of the promiscuous use of chemicals were suspected at least from 1962 when Rachel Carson published *Silent Spring*. And the first warning of impending climate change was given to Lyndon Johnson in 1965. But a half century later we still have no national climate policy and CO_2 continues to accumulate in the atmosphere faster than ever before.

The consequences of pollution gyres were not understood except in hindsight. In Wendell Berry's words, "we did not know what we were doing because we did not know what we were undoing." Even so, we knew better. And long ago we knew we had good alternatives such as recycling, energy efficiency, solar technology, and natural systems agriculture that have improved greatly in the years since. But widespread adoption of these alternatives was blocked by money, political dysfunction, and a lack of imagination. As a result, it is profitable for some to create a throwaway economy. It is highly profitable to extract, sell, and burn fossil fuels that are diminishing the human future. It is profitable to pollute our air, food, and water and undermine human health. The three gyres, in other words, are neither accidents nor anomalies, but the logical results of a system of ideas and philosophy deeply embedded in our culture, politics, economy, technology, and educational system.

The causes of the three gyres were once thought to be evidence of prosperity measured as economic growth. But a large part of our wealth is fraudulent. We are simply offloading the costs of pollution and environmental damages onto people living somewhere else or at some later time. We are beneficiaries of self-deception and conveniently bad bookkeeping.[4]

By undermining ecological balance, climate stability, and our reproductive potential, the three gyres are the primary causes of the "sixth great extinction" now under way. This time, however, it is not about dinosaurs and pterodactyls, but us. The approach path to oblivion, in Jean-Pierre Dupuy's words, is a "system of disruptions, discontinuities, and basic structural changes . . . feeding on one another and growing in strength . . . [leading to] an age of unprecedented violence." The stakes, in other words, are total, but there are no effective legal sanctions for the destruction of oceans, ecosystems, climate stability, human health, or actions that risk civilization for a few more years of corporate profits. We have yet to protect our descendants' rights to "life, liberty, and property." Neither do we acknowledge the right to life of our co-passengers on spaceship Earth. Our courts are blind to the plight of those who are suffering and many more who will assuredly suffer because of our dereliction. Indeed, there is no national or international legal regime commensurate with the human predicament or the requirements for ecological justice across generations.[5]

Most important, if one traces the causes of each gyre back in time there are students in classrooms acquiring the skills and mindset necessary to work unperturbed in the extractive economy that drives each gyre. Without knowing it, they are the dutiful acolytes of Descartes, Bacon, Galileo, and all of those in our time who share the dream of total human mastery over nature. We educators have equipped our graduates with the tools and technology necessary to enlarge the human empire, but not the wisdom to understand the consequences of doing so. Accordingly, generations of students have learned how to dismantle the world and concoct all manner of things but not why that was often a bad idea or how to repair the resulting damage. We have taught them how to manipulate, make, conjure, communicate worldwide, and sell everything under the sun but not how to think about the effects on themselves and others of doing such things. We have trained armies of lawyers and lobbyists with the skills to defend their right to plunder, but taught them nothing about enlarging the empire of justice across generational and species lines. We have taught the future leaders of mighty corporations how to grow their companies beyond imagination, but given them no guidance regarding the physical, ecological, and moral limits to the scale of the human estate or the concepts of enough and sufficiency.

Professor Tom Lewis captures the problem relative to the construction of the interstate highway system:

> The curriculum at engineering schools was extremely specialized and profoundly limited in its attention to the humanities and social sciences. Though they were charged with executing the largest civil engineering project in the history of the world, the students learned next to nothing about the effect their actions would have upon millions of citizens.[6]

Their teachers, in short, failed to provide a context for what they were teaching. The result was a highly skilled group of engineers unaware of and emotionally removed from the ethical, ecological, cultural, social, and international implications of their work. They were taught to be technicians, not thinkers, in a culture that is long on know-how and short on know-why.

The story could be an epitaph for Western culture. In large measure it is a result of an educational system in which students learn more than they can comprehend in ethical or ecological terms. Learning is a fast process, but comprehending the limits and proper uses of knowledge, which is to say acquiring wisdom, takes much longer.

My point is that the gyres of disintegration are not the work of the uneducated but rather the labor of those certified with PhDs, MBAs, LLBs, master's, and bachelor's degrees. In other words, the ecological and climate disorder we see around us reflects a prior disorder in how we think and what we think about. That makes it the business of all of us in the "education industry" who purport to improve thinking. But to improve thinking we must address problems *of* education not merely those *in* education, and so transcend the industrial-technological model of learning. Tinkering at the margins won't do.

The irony, of course, is that the same education, science, and technology that threaten life on Earth also gave us the capacity to discern the effects of our actions. We can measure our pollution down to parts per billion. We can chart the carbon dioxide accumulating in the atmosphere with great precision and take the temperature of the Earth with accuracy. We understand in detail many of the biological effects of long-term exposure to toxic substances. And since we know what we are doing, we can also decide to do better.

In the long view of history, however, we do not know yet whether the Western model of higher education will prove—on balance—to be a positive force in the evolution of a humane and sustainable civilization or simply a training ground for advanced cleverness serving an ever more powerful and destructive domination of Earth. If education is to play a positive role in a "Great Turning" toward a sustainable global civilization, our goal must be to enable coming generations to connect learning with a reverence for life and equip them with the analytical, practical, and emotional skills to be competent and caring stewards of the ecosphere. But that is hard to do in the blizzard of euphoria about our technological prowess and "breakthroughs" in every area except those that matter. It is harder to do when ideas and communication are being compressed into 140-character tweets that exist like flotsam in a flood of meaningless, decontextualized information. The difficulty is compounded by the fact that the rising generation spends on average nine hours a day in front of one kind of screen or another, in danger, as Hannah Arendt once said, of becoming "thoughtless creatures at the mercy of every gadget which is technically possible." A half-century later, Susan Jacoby lamented the "new age of unreason" in which "anti-rationalism and anti-intellectualism flourish in a mix that includes addiction to infotainment, every form of superstition and credulity, and an educational system that does a poor job of teaching not only basic skills but the logic underlying those skills."[7]

The fault, however, goes well beyond schooling. It has deeper cultural roots including the pathology that Richard Louv calls "nature deficit disorder." Since the dawn of the age of television, young people have increasingly lived indoors marinating in an entirely human-made world. The resulting damages are many: to the growth of intellect, to their sense of reality, to their basic affiliations, and to what biologist E. O. Wilson calls the "psychic thread" that connects us to nature. Louv argues that "the re-naturing of everyday life can be an important component of strengthening physical, psychological, and intellectual fitness . . . and relations between parents, children, and grandparents." Experience and mountains of data show that the emotional disposition to learn is enhanced by time spent out of doors and by the acquisition of practical skills.[8]

The same holds true in the design of educational buildings. People of all ages learn better and work more joyfully in places with sunlight, plants, "white" sound, fresh air, and natural materials. This is not a particularly astonishing revelation, merely an acknowledgment that we do better in settings similar to those in which we evolved. Design that integrates light, plants and trees, water, wind, rocks, and animals into the architecture of classrooms, buildings, and landscapes is not a luxury but rather a necessity, like coming home to a world we know in our bones.

The deeper challenge, however, is to transform the substance and process of education, beginning with the urgent need to prepare the rising generation—as best we are able—for a rapidly destabilizing ecosphere for which we have no precedent. We cannot know what they will need to know or how they should be taught, but we do know that they will need the kind of education that enables them to see across old boundaries of disciplines, geography, nationality, ethnicity, religion, and time. They will have to be intellectually agile without losing their sense of place and rootedness. They will need to rise above fundamentalisms of all kinds, including those rooted in the faith that more and better gadgets or an ever-growing economy can save us. They will need an ethical foundation oriented to the protection of life and the rights of generations to come. They will have to rediscover old truths and what biochemist Erwin Chargaff called "forgotten knowledge." They will need to know how to connect disparate fields of knowledge, how to "solve for pattern," how to design systems of solutions that multiply by positive feedback and synergy. We must educate them to be the designers of another kind of gyre that turns vicious cycles into virtuous cycles that might someday transform our politics, economy, cities, buildings, infrastructure, landscapes, transportation, agriculture, and technologies, as well as our hearts and minds. We need a generation that rises above despair or fantastical thinking and sees the world as a network of systems, patterns, and possibilities and so can give hope an authentic foundation.[9]

In other words, if higher education is to serve the interests of humankind and life in the long emergency ahead, it must be transformed, starting with a change in our concept of education and the purposes that ideas and disciplined knowledge serve. Samuel Johnson once said that the assurance of the gallows in a fortnight could concentrate the mind wonderfully. Similarly, the prospect of a civilizational collapse ought to concentrate our thinking about the substance and process of education in what could otherwise be "our final hour." We should better understand how we came to this point and what we can do to avert the worst that could be ahead.

Critics, predictably, will argue that saving the Earth, or humans for that matter, is not the business of educators, while refusing to say exactly whose business it is. Purists will argue that doing so involves making value judgments and the academy ought to be value free, which is itself a value and conveniently obeisant to the forces driving us toward oblivion. Pessimists will argue that transforming the academy is a good idea, but is not feasible and so should not be tried. Trustees will wish not to offend the powerful and wealthy and thereby risk one form of insolvency while presumably avoiding another. Incrementalists, for their part, will recommend caution and piecemeal change and hope that it doesn't come up a day late and a dollar short. And traditionalists, eyes to the rear, will want no change whatsoever.

But we no longer have the luxury of preserving the status quo. The landscape of education, including that wrought by the avalanche of television and electronic media, is rapidly changing and with it the mindscape of our civilization. Distance learning, online courses, and vendors other than colleges and universities are already redefining the content, process, economics, and meaning of education. What they mostly offer are cheaply acquired skills and quick credentials fitted for a turbulent job market in a devil-take-the-hindmost economy: mental junk food, not quality nutrition. They do not, and cannot, offer the depth and breadth of a genuinely liberal education that will always be labor-intensive and reflective, and can only work well at the pace at which prepared minds can comprehend complex, large, and portentous ideas. The fast educators have neither the time nor the inclination to facilitate deep consideration of education or enable their students to think about the act of thinking or offer useful guidance on the larger issues of our time. Their model of education is the business plan, their metric is cash flow, their students are just customers, their teachers are increasingly underpaid and overexploited adjuncts, and their pedagogy is that of the assembly line.

Consequently, the only reasonable educational response to the crisis of our age is a more thoroughly liberal and liberating education. Theologian Edward Long explains:

> A world without colleges and universities would be a place of contracting possibilities and shrinking horizons . . . [they] are the custodians of those intellectual resources without which this world would be engulfed in intellectual darkness and its inhabitants culturally disposed . . . A world deprived of its scholars and scholarship would be a world shorn of the most likely possibility of sensing the import of its history, a world without the tools to approach the present with perspective and a world of immediate preoccupations lacking a vision for the future.[10]

Such a vision must be grounded in the recognition that the gyres of dissolution begin with confusion and disconnection in our thinking and that all education, in one way or another, has to do with our terms of engagement with the ecosphere. By what is included or excluded we teach students that they are part of or apart from life and from its very creation. We need a consistent method and process for integrating knowledge and differing cultural perspectives into an ecologically coherent worldview. As Michael Crow observes:

> The academy remains unwilling to fully embrace the multiple ways of thinking, the different disciplinary cultures, orientations, and approaches to solving problems that have arisen through hundreds if not thousands of years of intellectual evolution. Our science remains culturally biased and isolated.[11]

Indeed, most colleges and universities have begun to reduce energy use, material waste, and adopt higher performance standards for buildings—what is called the "greening of higher education." The movement has grown throughout higher education because it was economically smart and also the "right thing to do." Yet the ecological and social assumptions in our courses, curricula, and research are hidden in preanalytic assumptions, and our conversations about curriculum changes are notable mostly because they are so trivial and forgettable. We argue a great deal over credit hours, parking, perquisites, salaries and benefits, and standards for promotion—anything but how our disciplines intersect with the fate of humankind on a deteriorating Earth. Nonetheless, I believe that only a truly liberal education can disenthrall us from the suicidal fantasies of our age and enable us to write a better story. In short, education can be part of the problem or the foundation for the great turning. But if education is going to lead the charge, what changes do we need to make now to the dominant educational paradigm and institutions of higher education?[12]

There are many possible answers but no single blueprint for change that fits the particularities of every college and university. For institutions committed to the free-flow of ideas, however, the starting point is a long-term, institution-wide conversation on the large questions of our time—the kind that have no easy answers and so would force us out of our comfort zones and beyond the conventional silos of thought and research. Since ideas and technologies have consequences, what is our responsibility for those that we let loose on the world? Are we complicit in the ecological disorder all around us? What are the ecological implications of our language and words like "resources," "progress," "natural capital," and "ecological services" that license human domination of nature? By what right do we justify colonizing the entire ecosphere and so causing the extinction of species and habitat destruction? How do we constrain our obsession "to predict, manage, and control nature"? Are there institutional, political, and moral limits to what we can know, or should know? Should we refuse to invest in or accept funding from, or perhaps divest from, certain companies or from entire industries? Should we impose a moratorium on things that have or could acquire the capacity to self-replicate? Similarly, ought we to supply the know-how for the transhumanist project to displace *Homo sapiens* with more intelligent machines that might someday find us stupid and fit only for menial labor? Similarly should we help to create synthetic life forms for which there is no evolutionary experience? Who should decide such things? If we were to attempt to restrain such endeavors, would that jeopardize other important values? And if we cannot place limits on research, how might we institutionalize precaution? In a tsunami of technological change, how can we maintain the ancient conversation about what it means to walk humbly with our God or even to live justly, responsibly, and well?[13]

Other questions will arise. What kind of knowledge will be necessary for the journey into the "Anthropocene"? What is the proper balance among intellect, heart, and hands? How do we join smartness with compassion? How should we improve the curriculum or reform the pedagogy to better prepare our students for the novel challenges they will surely face? How do we engage the humanities, social sciences, and natural sciences in ways commensurate with climate destabilization? How do we sustain our morale or that of students in difficult times and keep authentic hope alive? How do we calibrate our concerns for justice and fairness with a remorseless and unrelenting biophysical reality?[14]

There are also practical questions having to do with our responsibilities to the communities in which we exist. What do we know that could be put to good use in developing durable economies based on renewable energy and local farm and food systems? What do we know about nurturing decent and fair communities? How should we spend and invest institutional assets locally to promote development that can be sustained?[15]

From such ongoing conversations many results are possible. I will suggest only two of the most obvious. The first is a requirement that no one should graduate from any college or university without a firm grasp of how the world works as physical system and why that is important for their lives. We would be embarrassed to graduate students who could neither read nor count. We should be mortified, then, to graduate students who are ecologically illiterate—clueless about the basics of ecology, energetics, and systems dynamics—the bedrock conditions for civilization and human life. Before they leave the university setting, they should also be taught the social, political, economic, and philosophical causes of our predicament and have mastered the ethical, analytical, and practical tools necessary to build a durable, resilient, and decent world. In short, we should equip them with the capacity to integrate disparate subjects and disciplines into a coherent and ecologically grounded worldview.[16]

My second suggestion applies to institutional operations and management. Colleges and universities should not only advance the education of others but also themselves continue to learn how the ecosphere works as a physical system and to become more responsible within that system. Peter Senge describes the result as a "learning organization," one that makes sustainability the normal and easy thing to do, that is, the default setting for how we educate, build, spend, invest, and plan. More than any other institution, colleges and universities have an obligation to preserve the habitability of the planet that their graduates will inherit. In practical terms, this requires creating incentives that reward the efficient use of energy, water, land, and materials; that deploy renewable energy; and that preserve environments. It also means changing the rules and procedures in order to operate the organization as an ecologically sustainable system over the long term. Learning organizations require accurate and timely information and feedback, as my colleague John Petersen has shown in developing campus-wide dashboards reporting energy and water use. Above all, learning organizations would engage students, faculty, and administrators as co-participants in an ongoing design process that creates institutions and communities powered entirely by renewable energy and discharging no waste.[17]

The process will take perseverance and courage, but it all starts with what Martin Luther King called "the fierce urgency of now": "There is such a thing as being too late. Procrastination is still the thief of time. Life often leaves us standing bare, naked and dejected with a lost opportunity. We may cry out desperately for time to pause in her passage, but time is deaf to every plea and rushes on. Over the bleached bones and jumbled residue of numerous civilizations are written the pathetic words: 'too late.'"

E. M. Forster's admonition to "only connect" belies the fact that we are already connected. The greatest discoveries of the twentieth century revealed that we are stitched together in more ways than we can possibly know.

- Despite all that divides us, we share 99.5 percent of our genes with our fellow humans, and 98 percent of our genes with our nearest kin, the great apes and bonobos;
- A full 90 percent of our dry body weight isn't actually us but is instead a rowdy congress of bacteria, viruses, and other hitchhikers;
- Our minds have evolved to mirror each other's feelings and to empathize with each other;[18]
- Every breath we take includes molecules once breathed by Socrates, Lao Tzu, Shakespeare, Sojourner Truth, and, for that matter, Idi Amin;
- We have an innate affinity for life that Harvard biologist E. O. Wilson calls "biophilia";

- All of us are made of stuff that was once in stars;
- Plants are linked in networks, communicate by chemical signals, and help each other in ways that resemble altruism;[19] and
- We are now connected globally as never before by social media, emails, and smartphones in a thickening web of communication and intelligence that was predicted long ago by the theologian and philosopher Teilhard de Chardin.[20]

In short, we are connected over time as a small part of the vast enterprise of life that stretches back 3.8 billion years and will extend into the future as far as our better nature, luck, and sunlight may allow. The problem is not to connect, but to recognize and act on the reality of our connectedness.

Forster's further observation—that our capacity to connect "all turns on affection"—sounds quaintly irrelevant. Affection is the antithesis of the calculating mind that we associate with rational economic behavior, shrewd career decisions, and the self-referential narcissism that infects the "I" generation. It is also alien to the design of the modern research university, which features rationality over kindness, individual success over service, and manipulation without reverence. It permits every thought under the sun except for those having to do with limits and restraint. Affection is complicated and paradoxical. It thrives, however, at the crossroads where enlightened self-interest, altruism, and foresight meet. Affection is born in compassion, empathy, and an enlarged sense of self. It acknowledges that nothing and no one can be an island complete in itself. Everything and everyone are connected to the mainland.[21]

Affection changes what we think is important, trivial, or dangerous. It changes the substance and process of learning. It honors diversity and shuns uniformity and assembly-line pedagogy. Affection would cause us to better tailor learning to the student's personhood and manner of learning. It would calm the franticness and hustle of contemporary education. Affection would help us to acquire the patience to see learning as a lifelong process, one not to be confused with formal schooling. Informed by affection we would not so easily confuse information with knowledge or rationality with reasonableness. A foundation of affection would help us understand that thinking is often overrated and intuition underappreciated, and that true learning cannot be certified by grades and degrees. A dose of affection might even help us to comprehend and mediate the evolutionary divisions between the right and left hemispheres of our own minds.[22]

Affection deals in wholes, including the parts that are inexplicable and mysterious. It engenders a "feeling for the organism," as Barbara McClintock once put it. It connects us to the creative, artistic, musical, humorous, intuitive, and empathic sides of ourselves—qualities that often thrive only in the interstices of formal curriculum and the back alleys of colleges and universities. Albert Einstein put it this way: "The intuitive mind is a sacred gift and the rational mind is a faithful servant. We have created a society that honors the servant and has forgotten the gift."

Affection, then, causes us to celebrate mystery and opens us to the sense of wonder. Beyond the facts, data, theories, and analysis that permeate the academy, the inexplicable remains. What we know is like a drop in an ocean, whereas what we don't know is the ocean. Deep knowledge is elusive, the most furtive of creatures. The fact is that we are infinitely more ignorant than we are smart and always will be. And that's okay. D. H. Lawrence captured the essence of the matter by observing that "water is two parts hydrogen, one part oxygen but there is a third thing that makes it water, and no one knows what that is." And it is likely that no one ever will.[23]

Affection permits us to be compassionate with our own imperfect self and the imperfections of others. Affection isn't reserved for the easy times. In a world of paradox, irony, and tragedy, affection moderates pretensions and punctures illusions. It is kind and forgiving. Clear-eyed affection helps us acquire what Spanish philosopher Miguel de Unamuno once called "the tragic sense of life," which is neither resigned nor gloomy but instead permits us to laugh at ourselves and each other. It is the court jester, the trickster, and the fool who have mastered the art and science by which we live joyfully in a flawed world and get on with the business at hand. This approach has allowed us to triumph over tragedy before and equips us to do so again.

Intelligence disciplined by affection is different than that driven by raw IQ, catlike curiosity, or the scrambling ambition to succeed. Philosopher Mary Midgley explains why "smartness is not enough":

> This cult of "intelligence" centers on the idea that human cleverness is the supreme value . . . [but] all around us, we can see people trying to solve by logical argument or by the acquiring of information, problems that can only be dealt with by a change of heart—a change of attitude and new policy and direction. But this is the last thing we try . . . In contemporary culture, the passionate, quasi-religious exaltation of our pure cognitive faculties is surely a defense mechanism against this awkward fact.[24]

Putting smartness on steroids only makes things worse. Affection, to the contrary, would cause us to acknowledge the limits of our intelligence and be suspicious of all of those super-glued certainties we have about one thing or another. Sometimes those certainties are true, sometimes they are partially true, sometimes they are true under some circumstances and at some times but not at others, sometimes they are irrelevant, and sometimes they are dead wrong. Seeing the difference is called discernment, judgment, or humility—traits that we typically learn by falling on our face and getting back up and into the fray again. Affection, then, might cause us to see the difference between being right, or ardently thinking ourselves to be right, and being genuinely right and effective at making the world a little better. Affection, in other words, might cause us to be a little less certain of our certainties but clearer about our convictions. It would help us to realize that good intentions, including our own, aren't nearly enough; sometimes they're the problem.

Affection wears a smile, can laugh at itself, has a reverse gear, and seldom moves at warp speed. It takes its time. It is kind and civil in the manner of Martin Luther King Jr., Mother Teresa, Nelson Mandela, and Albert Schweitzer. It is never disrespectful.

Finally, affection helps us to see what could be without losing sight of how things are. Affection causes us to hope for improvement. And hope is a verb with its sleeves rolled up as something we do in daily practice, not just something that we wish for or talk about. It is a discipline requiring skill, competence, steadiness, and courage. It is practical. It bonds us to each other, and to real places, animals, trees, waters, and landscapes. The hopeful are patient, not passive. They are creators of the gyres of positive change that could, in time, redeem the human prospect. They are people who will know how to connect us to better possibilities waiting to be born.

eight
HEART

Philanthropy: Gk philanthropia 1. Love of humankind; the disposition or effort to promote the happiness and well-being of one's fellow people; practical benevolence.

—*Oxford English Dictionary*

Philanthropy is almost the only virtue which is sufficiently appreciated by mankind.

—Henry David Thoreau

There are nothing but gifts on this poor, poor Earth.

—Czesław Miłosz

In March 1982 William Zanker, the founder of the Learning Annex in New York City, announced that he would drop $10,000 in small bills from the 86th floor of the Empire State Building. He intended to express his appreciation to New Yorkers for supporting his adult-education school and get a bit of publicity in the process. But as fate would have it, twenty-eight-year-old Eddie Jewel and his colleague Salo Bandes, age thirty-four, picked the same day and hour to rob the bank on the first floor of the building. Thievery and philanthropy thereby collided but not necessarily as strangers to each other. The robbers struck first and were chased up 34th Street by plainclothes policemen accompanied by the gaggle of photographers intending to document a giveaway but now seizing the superior opportunity to record a getaway. A few minutes later Mr. Zanker and a colleague arrived at the scene carrying $10,000 in five clear plastic bags. In the confusion about which money had been stolen from the bank and which was to be floated down upon the masses, they were denied admittance to the tower elevators but were pursued about the lobby "by people snatching at the bags for samples." For his part, the building manager solemnly proclaimed that "the Empire State Building does not condone the dropping of dollar bills from its observatory this afternoon or anytime," a prudent policy rigorously obeyed ever since. As the *New York Times* reported, the amateur philanthropist departed the scene in a patrol car, announcing that he would study "other means of sharing their profits." Philanthropy, "the love of humankind," alas, is not always easy at the operational level.[1]

In fact, good philanthropic intentions produce good results less often than one might expect. Intending to head off mass starvation, for example, the Rockefeller Foundation launched the "Green Revolution" in 1941. In hindsight, however, it is evident that it is not possible, ever, to do just one thing. In this instance the other things—mostly unanticipated or ignored—included displacement of large numbers of peasant farmers, which disrupted ancient cultural patterns, urban migration, the drawdown of ground water, overreliance on fossil fuels and chemicals, water pollution, corporate domination, and so forth. The revolution manifested as a mixture of blessing and damnation that continues to this day.[2]

Much of the same can be said about the exuberant "philanthrocapitalists" to whom it has been revealed that the talent and good luck necessary to acquire a fortune also bestow the wisdom to solve complicated social problems, mostly by deployment of the same technology that, as luck would have it, is the source of their own considerable wealth. By most accounts the results of their philanthropic endeavors have ranged from microscopic to disastrous, but the photo-ops with billionaires and smiling disadvantaged children are always ever so touching. Facebook founder Mark Zuckerberg's foray into the Newark school system with Governor Chris Christie and Mayor Cory Booker is a textbook case of parachuting money into a disaster zone for maximum political effect followed by a quick exit. The politicos and philanthrocapitalists, it seems, had higher priorities than helping kids. To his credit, Zuckerberg is reported to be making a second go of it in the San Francisco Bay Area schools. One could tell many similar tales of noble philanthropic intentions gone south. Philanthropy, it seems, has a habit of consorting with irony, paradox, and unintended consequences. But hope that it will be different next time springs eternal in proportion to the fortune available.[3]

The act of giving, in fact, is one of the most complicated of human relationships. Giving is easy. Giving well to good effect, however, is much harder. The art of giving the right thing, to the right organization or person, in the proper amount, and at the right time is an art form that requires judgment, wisdom, and luck. Does the gift meet a particular need? Is the need frivolous? Would it be better for the intended recipient to supply the need on his or her own? Would it be better to give later? Or give more or less? Can results be measured? What about important qualities that defy measurement? The questions are many. Underlying all of the uncertainty is the fact that we cannot know for sure what any gift will cause in the world and what it will do for the giver or for the recipient. The only certainty is that we cannot rightfully avoid the subject that permeates nearly every human relationship at every scale.

Through much of human history "the debt we owe to the Earth was remembered," as Margaret Atwood writes. "Each religion paid tribute to the sacredness of the Earth and acknowledged with gratitude that everything people ate, drank, and breathed came from it through providence. Unless people treated the gifts given by the natural world with respect, and refrained from wastefulness and greed, divine displeasure would follow, signaled by drought, disease, and famine." Giving was similarly understood as part of a system of reciprocity in which the honor of the giver and recipient were mutually engaged. It was a way to keep the peace both within and between social groups, and a gift was not a possession to be hoarded, but rather to be passed on to someone else. The purpose, however, went beyond giver and receiver. "A gift that does nothing to enhance solidarity," Mary Douglas explains, "is a contradiction." It is a total system in that every item of status or of spiritual or material possession is implicated for everyone in the whole community." In this older comprehension, gifts had soul accompanied by their history as they moved from hand to hand and were embedded in societies that took pleasure in public giving, generous expenditure, hospitality, and festival. The ecology of giving, therefore, requires that gifts not be hoarded, but passed on expanded and improved. The "gift must move" as Lewis Hyde explained in his classic book *The Gift*. "When we see that we are actors in natural cycles," he writes, "we understand that what nature gives to us is influenced by what we give to nature . . . we come to see ourselves as one part of a large self-regulating system." Capitalism, on the contrary, presumes scarcity and so rewards private accumulation, that is to say, hoarding. But most "primitive" cultures had no word for scarcity. In Hyde's words: "when the market moves mostly for profit and the dominant myth is not 'to possess is to give' but 'the fittest survive,' then wealth will lose its motion and gather in isolated pools . . . and wealth can become scarce even as it increases." Transactions thereby become commodity exchanges by which goods and services are rendered for a price. In the true gift relationship, on the contrary, people bind themselves in a community of perpetual obligation. Some proffered gifts must be refused because the relationship, hence the obligation, is wrong or evil. The gift relationship works best at a relatively small scale because "our feelings close down when the numbers get too big." And that, according to Hyde, is "an unsolved dilemma of the modern world . . . in which empires of commodity expanded without limit until . . . all things—from land and labor to erotic life, religion and culture—were bought and sold like shoes."[4]

In the older sense, then, the purpose of giving was social cohesion, not necessarily benevolence, or even the usefulness or value of the particular gift. The purpose, as Hyde and Douglas explain, was not utility or even delight as much as order and conflict avoidance. Gifting was a means to the larger goal of holding things together and maintaining the peace. Times have changed. For the most part, we do not live in tribes and small communities, but in vast, sprawling metroplexes. Our relationships are mediated by the abstraction of the market, pervasive advertising, and invasive technologies that have released a flood of commodities and commercial temptations untethered from reciprocity. Calculation, individualism, and materialism instead prevail over the older notion of community cohesion. Gifts as purchased commodities are often trivial things devoid of any larger meaning or communal purpose. They are given as temporary diversions or status symbols with little more thought than, say, dropping dollar bills from the 86th floor of the Empire State Building. In fact, it can be difficult to know exactly what one causes by giving. This is particularly true for Western cultures because giving is presumed to be one-way, one-off transactions between individuals with no particular communal purpose, enduring obligation, or mandatory reciprocity.

Even stripped of its ancient context and communal obligations, the gift relationship has become more complex and paradoxical. In giving, one assumes that the gift fits a real need—that it is not superfluous, burdensome, or likely to result in unintended and undesired consequences. For example, giving the keys to a high-powered sports car to a teen-age boy can lead to trouble because the testosterone/horsepower ratio is all wrong. The gift can quickly morph into an unintended but somewhat predictable taking while also depriving the receiver of the opportunity to learn self-discipline and the gift inherent in the larger lesson that one ought to earn one's way in the world. Discerning one's own needs requires a self-knowledge that receivers may not have. Thoughtful givers, then, must have judgment, perspective, and wisdom. Moreover, giving ought to be premised on the assumption that the receiver can put the gift to good use. The gift of a simple blessing, say that given mistakenly by Isaac to Jacob who had disguised himself as his elder brother, Esau, became a curse because of the deceit involved. Shakespeare has King Lear dividing his realm between two daughters, a gift that became the source of tragedy because he misread their character and capacity to receive as givers with wisdom and charity.

Giving can be a form of entrapment as well. The Trojan horse that the Greeks bestowed on the people of Troy was malice masquerading as a gift. Along the same line, Dostoyevsky's Grand Inquisitor says of the masses: "Oh, never, never can they feed themselves. In the end they will lay their freedom at our feet and say 'make us your slaves, but feed us.'" The gift of food was no gift at all, but rather a taking meant to control the recipients and given in utter contempt of their rights, dignity, and potentials. By contrast, a true gift wisely given improves possibilities and helps make the recipient and perhaps the world a bit better.

The relation between giver and receiver, then, involves a blurring of roles whereby the act of giving can result in entrapment of one kind or another or can create dependency, intended or not, on the giver. In returning gifts received from Hamlet, Shakespeare has Ophelia say, "Rich gifts wax poor when givers prove unkind." But one can also give in such a way that the gratitude of the recipient is its own return gift. A child's innocent delight in a new toy, for instance, can be a gift of sorts given back to the parent—one that brings a delight of its own. Even without ancient communal obligations, there is no escaping the human and psychological complexities of the gift relationship in the fact that giving and receiving occur on both sides of every gift transaction. One gives selfishly, as a taker, or with sincere generosity, as a giver, while another receives as a taker or giver and is thereby obliged to return the gift in some fashion to others at some future time.

We also cannot escape the fact that giving and receiving define us as persons, as institutions, as economies, as societies, as nations, and, as we are beginning to realize, as a species. At each level, we are the sum of what we give and what we take. Perhaps in some cosmic bookkeeping office balances are kept and due bills or credit slips are issued from time to time.[5]

The larger scale of modern life has changed the way that gifts are given and received and how we calculate the relative costs and benefits on both sides of the transaction. Lewis Hyde considers this an "unsolved dilemma," that is, "how we are to reserve true community in a mass society, one whose dominant value is exchange value and whose morality has been codified into law." The fact is that we aren't very adept at accurately calculating at this scale, mostly because so far we have not cared to and even if we did, the calculations are difficult to make and always disputable.[6]

But modern societies give, nonetheless, perhaps more than ever. Benevolence flows through large-scale government-sponsored programs such as welfare, Social Security, and Medicare—but whether such programs are a net benefit or a net loss to society is a matter of debate between liberals and conservatives. In either case it is, as Mauss says, "still filled with religious elements." Internationally, we give foreign aid, disaster relief, and military assistance, believing these to be proof of our boundless national generosity. The Marshall Plan of 1947 is frequently cited as a noble act of benevolence, though it was in fact animated by our national self-interest in countering the suspected expansionist plans of the Soviet Union and in creating markets for American-made products. It was neither pure benevolence nor a deceptive taking, but instead a smart policy of the sort that we now call "win-win" that extended American influence, restored economic stability in Western Europe, and made a great deal of money for some in the process. In recent years, much of our foreign aid either distributes weapons to client countries or requires the recipients to purchase American goods, services, and agricultural products. This is not benevolence at all but a means to create and maintain dependent client states and promote corporate profits, yet another form of dependency, that is to say, a taking. Nonetheless, it is requisite that such giving fits the national myth of American goodness while serving the not-so-hidden goal of American preeminence.[7]

Closer to home, in 2014, $54 billion was doled out by some seventy thousand foundations ranging in size from the Gates Foundation with its $41 billion endowment in 2013 to small family foundations with a few thousand dollars. Their considerable assets are untaxed so long as they are used to benefit society, but it is left mostly to the donors to define the word "benefit," a sizeable zone of great but unquestioned imagination. Foundations and their trustees, officers, and staff are the gatekeepers who control access to a vast pool of money said to be some $700 billion, which is sheltered from taxation because it is presumed to be better spent for purposes deemed public by those who made the money in the first place. But foundations have their critics.[8]

According to Joel Fleishman, a former foundation president, foundations are "the least accountable major institutions in America . . . operat[ing] within an insulated culture that tolerates an inappropriate level of secrecy and even arrogance in their treatment of grant-seekers, grant-receivers, the wider civic sector, and the public officials charged with oversight." Their "besetting sins" include arrogance, discourtesy, inaccessibility, arbitrariness, failure to communicate, and "extreme attention deficit disorder." Many agree with Fleishman's views. Given the opportunity, they might propose remedies beginning with the stipulation that no one should be permitted to preside over the granting process who has not first been a supplicant and learned the hard way about the value of empathy, humility, returning phone calls, and responding to emails and who has perhaps worn out a few knee pads and hat brims in the process.[9]

More to the point, grants from foundation endowments, which would otherwise be taxable and used for public purposes, are seldom given for programs that run contrary to the rules of the system by which it was accumulated. In other words, foundations, more often than not, are allergic to controversy and risk. Contentious issues such as climate change and its deeper causes have been ignored or deferred even decades after the science documenting the scale of the danger was settled. With a few exceptions, then, the response of foundations to the enormous problem of climate destabilization has thus far been Lilliputian. In 2014 less than 2 percent of foundation grants went toward slowing or reversing climate destabilization. Gara Lamarche, president of the Democracy Alliance and previously the director of U.S. programs for the Soros Open Society Institute, explains:

> Courageous risk-taking is not what most people associate with foundations, whose boards and senior leadership are often dominated by establishment types. If tax preference is meant primarily to encourage boldness, it doesn't seem to be working . . . Too much of what philanthropy does is a sideshow to the most pressing issues of our time.

> Global warming threatens the planet . . . yet the handful of founda-
> tions engaged on the issue of climate change are vastly outgunned,
> and even they hoard their foundation assets as if there was no
> tomorrow—which in fact, there won't be before long if we don't re-
> verse the warming of the planet.[10]

The reason that climate destabilization does not attract foundation attention is that the great majority of trustees, foundation administrators, and program officers—even at this late hour—know
relatively little about climate science and how the Earth works as a
physical system and why such things matter for what they and their
grantees do. They are selected or hired because they have been
successful in other endeavors, notably finance, business, law, media,
academics, and public affairs, fields that offer little incentive or occasion for serious reflection on such things as the fate of the Earth. Not
having thought much about it, and consorting with others similarly
disinclined, they have not thus far been moved to do much about it.
With some exceptions, they are, by and large, not people easily alarmed
even by alarming things and so are inclined to overlook rapid climate
change as only another item on a long list of problems. In philanthropic circles, then, climate destabilization lacks priority among the
myriad of other concerns.[11]

A second problem is that climate destabilization crosses virtually
every boundary, physical and analytical, and so defies our modern
predisposition, in philanthropy and elsewhere, to organize the world
by pigeonholes, bureaucracies, silos, disciplines, and programs. It
also requires a broad and inclusive systems view of the world and a
time horizon that stretches over centuries—even though foundations
and indeed most organizations tend to privilege the short-term, immediate problems and disregard those farther down the road. The
result is that the stability and habitability of the Earth itself is regarded as a problem that can be solved by doing more of what we
have been doing all along, much of which is in fact culpable. With
few exceptions, foundations have been loath to confront deeper questions about the origins of our situation, perhaps because they are

inconveniently rooted in our current ways of living, working, thinking, and governing. Consequently, even "progressive" foundations have been mostly content to fiddle at the edges of problems, and avoid the underlying causes. The result is a lack of analytical and moral clarity about an issue that threatens the stability and very existence of civilization, not in some distant future, but much sooner.[12]

Somewhat ironically, defenders of foundations list among their great strengths their power to offset overbearing governments and their ability, indeed obligation, to operate on a longer time horizon than governments and businesses. "Foundations," in the words of Rob Reich, "are thus especially well placed to fund public goods that are under-produced, or not produced at all, by the marketplace or the state." Foundations can "take on the long-run, high-risk policy experiments that no one else will." These are not small levers for change. The question is whether there is the will to pull them.[13]

Until very recently, and through almost all of human history, each generation left things more or less as they found them. Rivers continued to flow seaward; animals, birds, and fish continued to thrive; forests grew lush and verdant; the seasons passed much as before; and the oceans faithfully yielded their bounty. From time to time someone would notice stones sticking up where fertile soils once existed or that forests preceded civilizations and deserts followed. But on the whole humans lived in an "empty," self-repairing world and if the fabric of life frayed in one place it was unimpaired elsewhere. The conventional wisdom, therefore, held that ecological losses were more than offset by the advances of civilization in the form of libraries, scientific knowledge, medical science, public health, culture, and increasing wealth. Each generation, accordingly, was presumed to be better off than the one before. By and large that was a plausible view until, say, the middle of the twentieth century. From that time on the relationship between the generations has changed dramatically, driven by exponential growth in population, economic output, and the use of energy. In a matter of decades the human prospect has radically changed. "The earth, our home," in Pope Francis's words, "is beginning to look more and more like an immense pile of filth." Indeed, the accumulating pollutants and trash, degraded landscapes, loss of species, deforestation, soil erosion, toxic wastes, depleted and acidic oceans, ubiquitous lethal weaponry, and worsening climate conditions are a legacy of debt and ruin imposed on future generations. A large fraction of the capital assets created over the last few generations was designed to accommodate sprawling, consumer-oriented, growth-obsessed societies powered by fossil fuels. That vast accumulation of roads, buildings, shopping malls, factories, and necrotic urban tissue is a burden that must be in large part rebuilt, dismantled, or abandoned. In other words, starting in the middle of the twentieth century we began to live as if there was no tomorrow—writing the preface to a self-fulfilling prophecy.[14]

In time and with sustained effort and ingenuity, some of the damage can be repaired. Forests can be replanted, some ecologies restored, and cities redesigned and rebuilt. But the atmosphere and oceans will take centuries to millennia to reach a new equilibrium, while the loss of species and changing coastlines cannot be reversed in a time span meaningful to us. In other words, the passage of a beautiful, bountiful, and fecund planet from one generation to the next can no longer be taken for granted. Unless we act with unprecedented speed and wisdom, our children and theirs will come to regard ugliness, ecological dysfunction, and increasingly capricious weather as normal. Baselines will shift downward. They will have no memory of the world that their ancestors once knew. In other words, unless we act now, the legacy of our generation will be a steadily worsening and unpayable ecological and spiritual debt.

What is now at stake is so immense that we hesitate to call it by its right name. Indeed it is so irrevocable, irreversible, vast, and unprecedented that we have no words for it. Call it the "death of birth," the end of evolution, "ecocide," or the destruction of civilization. But by any name we have no adequate way to describe the enormity of what is at risk and the losses and suffering that we are imposing on posterity.

But how, in the age of science and information, did this come to be? Of many possibilities I will mention the two that I think will prove to be helpful in making the transition to a decent future. The first is that along the way to the postmodern world, we lost much of the capacity for gratitude and ceased to believe that the mystery of life was a gift at all. Life happened by chance and evolution and not as a gift that implies a giver or the obligation to return the gift. In a more calculating time, gratitude ran inversely proportional to favors not yet granted. It was not always so. For Seneca and Cicero, for example, ingratitude was among the worst of human faults. Dante places ingrates in the ninth circle of Hell where they are eternally frozen, prostrate in postures of deference. Shakespeare, too, loathed ingratitude "as treason and faithlessness, as the source of every kind of wickedness." And Goethe regarded ingratitude as "a kind of weakness. I have never known competent people to be ungrateful."[15]

In a more secular and scientific time, however, we have dispensed with rituals and celebrations by which we once participated in the mysteries of seasons, fertility, and the business of appeasing irritable and demanding deities. Our superstitions now serve other ends. As Margaret Visser writes, "Gratitude takes time . . . thoughtful involvement with other people. But we live in speed-driven, unthinking separation from one another, each ignoring the interests of others in order to pursue his or her own." Furthermore, "we have cut back on the number of occasions where gratitude can arise or be considered an appropriate response." Money and self-interest, narrowly conceived, trump other values.[16]

When required by ceremony or self-interest, gratitude is uttered in the desiccated language of self-interest or in the various dialects of utilitarianism. Life and nature are no longer regarded as gifts; instead they are considered economic assets. Somewhere along the way, we began to measure forests in board feet, water in acre feet, ancient sunlight as BTUs and kilowatt hours, ecologies as biomass, and our own work as man-hours. We have measured, managed, and grown rich. We "are remodeling the Alhambra with a steam shovel," as Aldo Leopold once put it, "and we are proud of our yardage." Outside of a small tribe of dreamy romantics, nature lovers, and tree huggers, we've been busy getting on, fighting wars, building empires of one kind or another, and growing our economies. It was, as William Catton once said, the "age of exuberance" and we opted out from restraints once imposed by religion, philosophy, courtesy, and societal bonds, as well as those imposed on us by rudimentary technologies. God, as Nietzsche once noticed, is dead. We saw little further reason for gratitude and little offense from ingratitude. After all, to whom or what should we be grateful? Writer Margaret Atwood compares our situation with that of Doctor Faustus:

> Mankind made a Faustian bargain . . . as soon as he invented his first technologies . . . now we have the most intricate system of gizmos the world has ever known. Our technological system is the mill that grinds out anything you wish to order up, but no one knows how to turn it off. The end result of a totally efficient technological exploitation of Nature would be a lifeless desert: all natural capital would be exhausted, having been devoured by the mills of production, and the resulting debt to Nature would be infinite. But long before then, payback time will come for Mankind.[17]

Payback will come first and hardest for those least responsible for the climate debacle and ecological impoverishment, but eventually it will spare no one except the generations that started the avalanche of ruin. Looking back at us—their ancestors—our descendants might well wonder why we, who seem otherwise so adept at calculation, did not "calculate the real costs of how we've been living." From their vantage point in an uglier, more capricious, and threadbare world they might conclude that the fossil-fuel generations were stupid, ignorant, heartless, or some combination of all three. In other words, we, the beneficiaries of fossil fuels, were derelict as trustees of the commons of air, water, climate stability, land, forests, biological diversity, human flourishing, and of life itself. We will be remembered as the ones who violated the rule that the gift must move, passed on in full to all of those who will follow or might otherwise have followed. Perhaps they will regard us rather like we regard disabled people, unable to overcome "the impoverishment of our systems of practical reason, the paralysis of our politics, and the limits of our cognitive and affective capacities." We have become blind to our own indebtedness and crippled by the demands of a commercial system in which "gifts" are merely commodities given for an ulterior purpose.

But there is an alternative. Margaret Visser describes an

> "anti-economy" that is the realm of the gift, people do not hang on to what they have, nor do they seek profit above all else. Instead, having recognized the extent to which they have been gifted themselves, they pay attention to the needs of others and recognize—look squarely [at]—not only . . . the problem but at those who are suffering injustice or are just plain suffering . . . they do not feel that what they have given away is something they have lost.[18]

The "anti-economy" is, one suspects, the most prosperous economy of all. It accounts for its true and full costs, it shares its abundance, it does not hoard. It is suspicious, but not always hostile to novelty. It is a means to things higher than itself. And it does not make a fetish of buying and selling.

For those who find concepts such as gratitude hopelessly naïve and vague, there is a second possibility that focuses attention on the failure of our laws and political institutions to protect the rights of posterity. The idea has a respectable history. In September of 1789 Thomas Jefferson wrote to James Madison, declaring that "the earth belongs in usufruct to the living; that the dead have neither powers nor rights over it." The key word is "usufruct," which the *Oxford English Dictionary* defines as "the right of enjoying the use of and income from another's property without destroying, damaging, or diminishing the property." His point was that no generation could rightfully encumber those coming after with debt, which he regarded as financial even though I suspect he had intuitions of other and larger debts as well.[19]

A year later Edmund Burke, a founder of modern conservatism, wrote that "our liberties [are] an *entailed inheritance* derived to us from our forefathers, and to be transmitted to our posterity; as an estate specially belonging to the people of this kingdom without any reference whatever to any other more general or prior right." He went on to describe society as a "contract . . . a partnership in all science; a partnership in all art; a partnership in every virtue, and in all perfection . . . a partnership not only between those who are living, but between those who are living, those who are to be born . . . [part of] the great primaeval [*sic*] contract of eternal society, linking the lower with the higher natures, connecting the visible and invisible world, according to a fixed compact sanctioned by the inviolable oath which holds all physical and all moral natures, each in their appointed place."[20]

A century earlier, the question of what one generation could take from others and appropriate from the Earth was also a concern to John Locke who, more than any other, informed the political thinking of the founding generation. In his *Two Treatises of Government*, Locke placed no limit on ownership of property properly improved by human labor except that "enough and as good [was] left in common for others." But how much could a man have? Locke's answer: only "as much as anyone can make use of to any advantage of life . . . whatever is beyond this is more than his share and belongs to others." Locke went further, asserting that "nothing was made by God for Man to spoil or destroy."[21]

From very different perspectives, then, Locke, Jefferson, and Burke agreed on the basic principles that no generation had the right to diminish the prospects of those to follow and that there are limits to what any one person or any one generation might rightfully accumulate at the expense of another. Theirs was a more measured and sober age, one informed by a common faith in the clockwork orderliness of things, whether bestowed by Isaac Newton or God. They could not have foreseen our present circumstances in which humans are disrupting climate stability and thereby jeopardizing those who could—or would—come after. They could not have foreseen the sheer scale of damage that humans would soon inflict on the whole of nature. They could not have predicted the onset of the sixth extinction or the continuing threat of nuclear holocaust. Nor could they have foreseen the possible unraveling of civilization. They were raised in an age that was both more optimistic and perhaps more realistic about human foibles and fallibilities. They were men of the eighteenth-century Enlightenment and as such were optimistic about many things, even if they were not necessarily sanguine about human nature. They were realists about the limits of humankind and about the power of governments to hold things together against the centripetal forces of faction and pettiness. They neglected, however, to protect an increasingly vulnerable posterity by defining a larger ground of rights robust enough to extend within and across generations and even species. That is the great and urgent work of our generation.

Economist Kenneth Boulding once jokingly asked, "What has posterity done for us lately?" The answer, of course, is "nothing." And in our age, which is so consumed with self-interest as the sole legitimate human motivation, the unspoken conclusion follows that we have no binding obligation to future generations, which are powerless to reciprocate. Despite occasional lip service to the contrary, our economy and our politics are calibrated to the short-term and neglect those who come after. But have we missed something? Are there reasons persuasive and powerful enough to override the perceived self-interest of entire generations that would compel them to leave "as much and as good" for subsequent generations? A great deal depends on how we answer that question, so let me offer two and a half possible answers. The first is drawn from the great conservative Edmund Burke, who wrote, "People will not look forward to posterity, who never look backward to their ancestors." In his view, a chain of obligation connects past, present, and future generations and when honored can help overcome the "selfish temper and confined views" of the present—that is, it can work in our self-interest. Boulding, similarly, argued for a broader concept of self-interest, saying in essence that indifference to our posterity was, in effect, the flip side of indifference to our own self-interest here and now. Carelessness, in other words, is highly fungible, whereas benevolence and foresight are indivisible across generations, so that the measures we take now to protect future generations will benefit those presently living.[22]

Many questions follow. The argument is plausible insofar as it pertains to the protection of natural capital of soils, biological diversity, and limits to climate change. But is it justifiable regarding current economic growth and efforts to raise the poor out of poverty? How far into the future should our concern extend? What is the tradeoff between that which is due to the living and that due to posterity? Who defines humankind's self-interest? And so forth. The fact is that what is taken for self-interest in one time and in one set of circumstances changes over time and when seen from a different vantage point. The historical record says that we are a fickle lot.[23]

A second reason to extend rights to future generations is traceable to the doctrine of natural law. Once thought to be derived from God, it is more plausible to assume that it is, as philosopher Leszek Kołakowski argues, "a man-made convention . . . a set of rules embedded in the ontic condition of humanity—in human dignity." In an age of science, however, may we still believe in something so undefined as natural law? Kołakowski argues strongly in the affirmative: "Not only may we believe in natural law, but by denying it we deny our humanity." The alternative is a kind of nihilism that leaves our better instincts bound, gagged, and helpless in the face of moral drift and outright evil. Natural law is premised not on the existence of God, but on what he calls "the moral constitution of being—a constitution that converges with the rule of Reason in the universe." The observance of natural law further "erects barriers that limit positive legislation and do not allow it to legalize attempts to infringe the indestructible dignity that is proper to every human being . . . invalidat[ing] slavery, torture, political censorship, inequality before the law, compulsory religious worship, or the prohibition of worship, or the duty to inform the authorities about the non-conformity of people's political views."[24]

Natural law is a close kin, I think, to what naturalist Aldo Leopold proposed as the "land ethic." "All ethics so far evolved," he wrote, "rest upon a single premise: that the individual is a member of a community of interdependent parts. His instincts prompt him to compete for his place in the community, but his ethics prompt him also to cooperate (perhaps in order that there may be a place to compete for)." The land ethic, for Leopold, "simply enlarges the boundaries of the community to include soils, waters, plants, and animals, or collectively: the land[,] and changed the role of *Homo sapiens* from conqueror of the land-community to plain member and citizen of it." Refusal to abide by the rules and limits of living as "a cog in an ecological mechanism" of the larger land community carried a penalty: "it would grind him to dust."[25]

The same remorseless logic underpins gift economies, in which "above givers and receivers stands a law, as merciless as it is abstract . . . if you receive a gift, you must later on reciprocate." Failure to do so incurred severe and unavoidable penalties. The presumed physics of natural law, the land ethic, and gift economies reflected assumptions about the deep nature of reality that can be neither proved nor disproved. But our common language often suggests affinity for something of this sort when we say things like "what goes round, comes around" or when we refer to "chickens coming home to roost." We can debate the origins of primal morality, whether implanted in us by God as C. S. Lewis thought, or whether by evolution from the bottom up as some biologists believe. God alone knows and he, she, or it isn't saying. (Just for fun, perhaps God made an occasional skeptical Brit like Richard Dawkins to keep things interesting.)[26]

In either case, we instinctively recoil from doing violence to our children and kin. We instinctively affiliate with "life and life-like processes," or what biologist E. O. Wilson calls "biophilia." Very young children have a rudimentary, prearticulate sense of fairness. I am inclined to think that compassion, kindness to strangers, mercy, and forgiveness are in some patchy way woven into our behavior. Deep in our bones, some things just feel right and others abhorrent. This, I think, is the substratum of our still-evolving moral consciousness and may explain why it seems very odd to debate whether—by any stretch of logic or sophistry—a few generations have the right to hog more than their fair share of Earth's resources, including climate resilience. We are the first and perhaps the last generation likely to have the moral elasticity and the inclination to debate such questions. No prior generation had the technological capacity to threaten more than a small sliver of the biosphere or significantly damage the human experiment when dispersed over six continents connected only by sailing ships. Assuming that humankind survives through the gauntlet ahead, our presumably chastened and wiser posterity will better know why life, the ecosphere, and civilization should never again be put in such peril for any reason whatsoever.

My half reason to act on behalf of future generations follows. It is discounted by half because it rests entirely on the unverifiable grounds of my own feelings and experience such as they are. I cannot say for certain whether my life is a gift or not, but even with its ups and downs it certainly feels like one. I cannot say for certain that the feeling of sea winds in my face, or the fragrance of spring flowers, or the sounds of whitewater, or the comforting solitude of an old-growth hemlock forest, or the view from an Appalachian ridgetop, or the feel of good oak, or the hug of an old friend, or the smell of rain after a long hot drought or that of newly plowed ground, or the red-tail hawk that is nesting in my yard this spring, or the delight of fireflies on a summer evening are gifts. But they give me pleasure and they feel like beneficence beyond any thoughts I can muster. I cannot say with cool scientific logic why such things should be passed on unimpaired to my four grandchildren. But I fervently hope that they will be. I can say only that I am very thankful that others before us protected what they did or were unable to damage more than they did.

Whether life and all that it entails, including pain and suffering, has come to each of us as a gift or by random chance no longer interests me. How could I know such things? Either way I can imagine no good reason to destroy, cheapen, deface, or risk harm to the biotic community and the enterprise of life. I can, however, think of many reasons to protect life from the vast planetary cycling of water and materials down to the pair of cardinals nesting this spring in a cedar tree in my side yard. I believe myself to be better off because they're busy raising their brood and that makes me happy. I am even willing to chase off the neighbor's cat prowling close by on their behalf. Exactly why I cannot say, nor do I think it very important to give reasons or analyze further. In all of our effort to conjure up reasons for one thing or another down to parts per million, one thing missing is the sense of wonder that exists beyond words and reasons. Theologian Abraham Heschel once said that "a life without wonder is not worth living." In that spirit, I am content to find my cardinals and their brood wonderful and feel blessed by their presence in some mysterious, unnamable way and let it go at that. But I am distraught by the possibility that these wondrous things may go extinct, never to be experienced by my descendants and yours.[27]

The paradox of a true community that includes people, cardinals, and generations of both yet to be born is that its individual members do better as the larger community flourishes and vice versa. We live as parts of larger systems, a fact that is both restraining and liberating but either way is irrevocable. Our puny rebellions on behalf of our individual rights and flag-waving, gun-toting, huffing and puffing confederacies always come back to hurt us and the communities in which we live. I am curmudgeon enough to think that the same will hold true of the other rebels: the narcissistic herd of solipsists of the iPhone, iPad, iGeneration who aim to secede from one thing or another yet wish to merge in the great mashpit of the singularity. For a long time to come we will argue over the rules for running communities and other systems, but for now Leopold's land ethic will do: "A thing is right when it tends to preserve the integrity, stability, and beauty of the biotic community. It is wrong when it tends otherwise." We may be debating that cryptic bit of eloquent wisdom for a long time. But it makes a point about our dependencies that is overlooked in our commercial society, which conceives of the world as a cornucopia of one-way flows, not as a system with feedback loops. In the face of that reality, reverence is the better part of discretion.

I opened this chapter with the *Oxford English Dictionary* definition of philanthropy as "the love of humankind; the disposition or effort to promote the happiness and well-being of one's fellow people." But it is not possible to truly love humankind apart from its ecological context. The two are one and indivisible but they are not symmetrical. We may claim to improve nature and in certain places I suppose that is true in a limited way, but on the whole nature has no obvious need of us. Any relatively nondestructive keystone species would do as well or better. In other words, our pretentions otherwise aside, we are entirely dispensable and unfortunately seem determined to become more so.

For serious philanthropists of all kinds, the first challenge of truly loving humankind requires that we understand, protect, and, when possible, enhance the natural systems that nourish our bodies and souls. Restoring health to the systems that we have damaged and on which we depend, however, is more than tidying up a bit after a binge. It is rather an act of atonement for the original sin of being so casually, carelessly, and sometimes wantonly destructive of things about which we knew so little and on which we depend so much. But it requires no more than an enlightened, ecologically informed self-interest. The second challenge is harder and goes further. It is to make certain that the good Earth is passed on in full to those who will (or would have) come after us. Call it a gift if you so choose, or trusteeship or stewardship, but by any name the safe passage of Earth to coming generations would be the first deliberate act of true philanthropy from one generation to another.

By the direst of dire necessities I stumbled into the world of philanthropy in the 1970s. In the years since I've been on both sides of the desk, serving as a trustee, adviser, or fellow to ten foundations and as a fundraiser for two organizations. I cannot say whether it is more blessed to give than to receive, but I know for a fact that it is a lot less nerve-wracking. Good fundraisers quickly learn to measure people by their cash value and by whom they know with possibly even deeper pockets. Like all careerists and courtiers, the successful ones learn the fine arts of solicitation, supplication, and, well, sucking up with style. They learn to accommodate the word "no," smile, and come back later. It requires a long view of possibilities and it does take a toll.

On the other side of the desk foundation program officers soon discover that their jokes are funnier than they used to be, their insights much deeper, and their suggestions are always incredibly helpful. They are in demand, invited to conferences in cool, exotic places to meet with such interesting people doing amazing and important work. They learn to sprinkle words like "strategic," "leverage," and "targeted" strategically into conversations with peers and in reports to trustees to gain maximum persuasive leverage to targeted effect and so forth. The airports get old after a while but the buzz is always cool. All the while they wear the attitude of easy assurance that comes with never having to worry about a paycheck, about having to live by their wits, or about having a cause for which they'll bet it all over a lifetime. That, too, exacts a price.

Giving is the easier part of it, but giving well to good effect is not easy. Done well, philanthropy requires good judgment, a lot of investigation, a gambler's instinct, a nose for opportunity, and the willingness to lose sometimes for the right reasons. Even so, it results in fewer sleepless nights. The president of a very large foundation once told me that his foundation had never given to a project or organization that had failed. He was proud of that fact, oblivious to the possibility that they'd apparently never risked much, either. Giving to big, well-established organizations is the way of the herd: it risks nothing and allows the donor to bask in the reflected glory of the rich and famous, which is not an inconsiderable thing.

A great deal of what I know about the better aspects of philanthropy I learned from people like Edith Muma, a longtime trustee of the Jessie Smith Noyes Foundation in New York. I first met Edith in 1987 when I requested a grant to fund the first-ever study of the food system on a college campus. Later I served for two terms on the Noyes Foundation board and witnessed the work of one of the most remarkable people I've ever known. Edith seldom said much during meetings until she was asked. When she did speak she spoke quietly, authoritatively, and with the kind of insight that changes perspectives around the table. Often she saw things that others had missed. She was compassionate but tough-minded. She could see large possibilities in small and seemingly insignificant beginnings. Her rule of thumb was to bet on people, and she was a shrewd judge of character, creativity, and stamina. Largely because of her, the Noyes Foundation funded early on many of the pioneers of the sustainability movement: Wes Jackson, Amory Lovins, Donella Meadows, and John and Nancy Todd. For me and for many others, Edith was a model of how philanthropists ought to nurture possibilities and those people attempting to nudge the world in better directions. She was a radical in the true sense of the word: she worked to get to the roots of what ails us. Edith understood the precariousness of our time and faced the prospects with a kind of gracious, compassionate, and quietly stubborn wisdom born of the faith that there are better possibilities. She was willing to bet on visionaries and long shots and more often than not they repaid her confidence many times over.

Edith Muma and her colleagues represented philanthropy at its best: purposeful, clear-headed, adventuresome, personable, and unpretentious. She returned phone calls, answered her mail, and did not impose on grantees. In a word, she cared carefully and artfully. The world is better for that. She embodied the true spirit of philanthropy: the love of humankind not as a theory but as applied love.

Nothing that is worth doing can be achieved in our lifetime; therefore we must be saved by hope. Nothing which is true or beautiful or good makes complete sense in any immediate context of history; therefore we must be saved by faith. Nothing we do, however virtuous, can be accomplished alone; therefore we are saved by love. No virtuous act is quite as virtuous from the standpoint of our friend or foe as it is from our standpoint. Therefore we must be saved by the final form of love which is forgiveness.

—Reinhold Niebuhr

Man went wrong when his brain outgrew his soul. In this era of big brains anything which can be done will be done—so hunker down.

—Martin Amis

It's late in the day. Shadows are lengthening, but on the outskirts of Lonesome Dove, the crowd is assembling for the migration to a better place called Montana. Most are disheveled, disoriented, lean, and hungry. Some are moderately well dressed, but appear confused and anxious. A very, very few are expensively attired, trim, tan, and fit. They are obviously accustomed to giving orders and having them carried out. Call them Davos men. They speak confidently about investments and new technologies and how they will make the journey profitable for a few and far easier for everyone else: solar power too cheap to meter, miracle materials lighter than feathers and stronger than steel, a new circular economy, an internet of things, driverless cars, smart houses that anticipate our every whim, robots to do our work, and much more. Theirs is a world of technological progress, economic growth, and opportunity; no cloud darkens their skies.

In the main, however, there is a confusing, mob-like, disorganized milling around with menacing whispers of discontent about inequalities, shortages, undue privilege, exploitation, and the fierce demands of one God or another. Off to the side are men in white lab coats huddled in rapt conversation while close by, human-looking creatures stand blank and aloof. The parking lot nearby has a few stretch limos with uniformed drivers waiting for their passengers; in the distance, a line of private jets with corporate logos idle on the tarmac. Off to the other side, menacing, purposeful, uniformed men wearing dark glasses stand near shrouded equipment. But maps, organization charts, a statement of principles, and contingency plans are nowhere to be seen. For all of their self-assurance, none of those who appear to be leaders have any idea of where Montana might be and no idea about what to do when they get there except more of the same things that had made Lonesome Dove so forlorn.

Metaphors, again, have their limits. Leaving issues of nuclear weapons and the threat posed by "superintelligence" aside for the moment, we know that the transition to environmental sustainability alone will be a very long process. It would be good to admit this at the outset and plan accordingly. Over that span of time we must first eliminate emissions of carbon dioxide and other heat trapping gases. We must stop the "sixth extinction" in its tracks. We must eliminate toxic contamination. We must reduce our "footprint" on Earth. We must conserve, preserve, and restore healthy natural systems. We must fairly and humanely bring population and consumption in line with ecological reality. We must create a far more just, fair, and peaceful civilization, otherwise all else will be moot. None of these "must do's" will be easy, though they will be easier than not doing them.

Like any journey, there are decisions to be made. What to take, what to leave behind? When hard times come along the way, who will make the decisions and by what means and by what standards? Will we govern by triage and leave the wounded behind? We are, in short, starting the largest, longest, most complex, and crucial transition in human history unprepared and disorganized. And it is beginning as a free for all.

Of course, there will be no single journey, but many different variations with different starting points and, eventually, different destinations. There is no such thing as once and for all sustainability but rather a continuing negotiation between humans and nature. No one knows how much time we have to repair and restore what we can of the planet's life-support systems, but we do know that we are starting late. The transition should have begun in earnest decades ago when the first scientific warnings of ecological stresses and limits were sounded. It should certainly have begun within a few years of the Stockholm Conference of 1972, or soon after Amory Lovins's proposal for a soft energy path in 1976, the publication of the *Global 2000* in 1980, or the release of *Our Common Future* in 1987. Had the United States led in those years, the transition ahead would be much shorter and smoother—not perfect, mind you, but more certain to put the world on a better path. Instead, Americans chose to believe that it was "morning in America again" and so embarked on a binge of profligate consumption, TV watching, and overeating. We indulged our animosities in phony culture wars fought mostly over what governments could or could not do in our bedrooms. At a considerable cost in lives (ours and theirs), and for no good reason, we stirred up monumental troubles in Iraq, a country that had plenty of its own. All the while we tolerated an astonishing level of political dereliction. But our recent history is partly the manifestation of our national schizophrenia about the relationship between society and nature, which runs like a geological fault line beneath the daily headlines. As a result, the transition to sustainability, however defined, will be a close call. There will be losses along the way and some won't make it. The generations living through the transition will be stressed in ways that we can scarcely imagine. And somehow, in what could turn into a pell-mell scramble, the cowboys from Lonesome Dove, born and bred in the exuberance of the fossil-fuel age, must become smarter, wiser, and less trigger happy.[1]

Some historical perspective will help. The United States led the sprint into the twenty-first century, but in fact we weren't ready for prime time for reasons rooted in our history and culture. Throughout the seventeenth century, science and technology began to advance, slowly at first and then with ever quickening speed. Henry Adams had a premonition that it could end badly. In 1862 he wrote:

> I firmly believe that before many centuries more, science will be the master of men. The engines he will have invented will be beyond his strength to control. Someday science may have the existence of mankind in its power, and the human race commit suicide, by blowing up the world.[2]

Four decades later, Adams speculated that a "Law of Acceleration" was driving history beyond our capacity to comprehend and control. Each advance led to many others. At first the effects were mostly positive: they improved human health and longevity, eliminated drudgery, increased wealth, and gradually improved the quality of life. The driving force was ingenuity unleashed from the restraints of ignorance, incompetence, culture, religion, and poverty, and finally wedded to government and corporate power. The sign above the entry to the 1933 World's Fair in Chicago said it all: "Science explores, Technology executes, Man conforms." Adams worried that the rate of technological change was outpacing our capacity to avoid the adverse consequences of our newfound powers. "Means have outgrown man," as historian Sigfried Giedion later put it. Lewis Mumford was even gloomier about the human prospect: "The union of automation and the id will probably bring about the catastrophic destruction of our civilization," he wrote in 1954. Václav Havel attributed our technological predicament to "the inevitable consequences of rationalism . . . [based on] a large and dangerous illusion: that no higher and darker powers ever existed, either in the human consciousness or in the mysterious universe."[3]

Much of the technological development they feared originated in weapons labs and appeared later in the marketplace as solutions looking for a problem. But whatever the historical causes and sources, we lacked the institutional and political capacity to comprehend and control the tsunami of technology washing over society. As a result, we neither anticipated nor accounted for the full costs, benefits, and risks of constant and frenetic innovation. With only a few exceptions (for example, supersonic transport and ozone depletion) we issued blanket exemptions from consequences under the guise of progress, free markets, economic growth, freedom to do as one pleased, or just that technological change was inevitable. We accepted "creative destruction" as a tolerable cost of capitalism and more recently "disruptive innovation" as a worthy goal for schools, public agencies, and corporations, without asking whether novelty is always an improvement or not; what exactly is being improved, or perhaps disrupted or destroyed; and why. The language of disruption, as historian Jill Lepore notes, is "a language of panic, fear, asymmetry, and disorder" suitable "for an age seized by terror." At the same time we rejected older ways of doing things that did not need much improvement and sped up things that for good reasons have a slower pace. Somewhere along the way we lost track of the differences between complexity and mere complicatedness, between should do and can do, and between real progress and rote change. A growing unease with these incongruities is evident in counter movements like the new urbanism, slow money, slow foods, sustainable farming, community forestry, and natural medicine. In various ways these and other efforts express disaffection with the drift of technology, the speed of change, and the loss of community. In other words, technological change with no speed limits jeopardizes what Simone Weil once described as the human "need for roots." It has also overwhelmed our politics, laws, and institutions and eroded ancient loyalties and affections. Living on adrenalin, we "overran our headlights."[4]

Here is what we know about the transition ahead. First, we have initiated long-term changes in the Earth's systems, including rapid warming, pollution, species extinction, population and economic growth, deforestation, and changes in land use. Changes are happening globally faster than scientists were predicting even a few years ago. As a result, "the living world," as E. O. Wilson writes, "is in desperate condition."[5] Second, the most urgent issue we face is climate destabilization, an "everything issue" and a symptom of deeper problems embedded in our politics, economy, and culture. Third, climate change and changes in the ecosphere are global; there are no places left to hide. Fourth, our time of troubles will last until climate and ecological systems reach a new equilibrium sometime in a distant future. By then the Earth could be a different kind of planet—Bill McKibben's "Eaarth." Fifth, our fate will be heavily influenced, but not necessarily determined, by the remorseless workings of large numbers having to do with Earth's biogeochemical cycles, the inevitable force of entropy, and processes of ecology that work at their own pace and without any consideration whatsoever for humankind. None of these offer leniency for good intentions or for good behavior performed too late. The transition ahead, in other words, requires the harmonization of the actions of humankind with the way the Earth works as a biophysical system. We will have to work without assurance of success but in the faith that there is time enough to avoid the worst and make the human presence on Earth sustainable and fair. Sixth, we know that there is no such thing as once and for all sustainability. At best there will be an ongoing dialogue between humans and the ecosphere, but under rules set by nature. God may forgive our sins, as Wes Jackson puts it, but nature won't. And there is no *Deus ex machina*, or cavalry, or invisible hand, or miracle technological breakthrough that will rescue us in the nick of time. It will be up to us to change the odds and the outcomes on our own, and that process, as Pope Francis reminds us, can begin only with a change of heart—however improbable that might seem to skeptics. Finally, other threats to our tenure on Earth exist at the periphery of our awareness. Nuclear weapons still pose a mortal threat to the human prospect. And as noted, there is a growing possibility that the momentum of artificial intelligence could usurp human agency, inaugurating a post-human future.[6]

Despite all of the complexity and unknowns, the first steps ahead are obvious, necessary, economically advantageous, practical—and insufficient. I refer to the transition to both energy efficiency and solar and wind energy, which is by most accounts moving rapidly because of their declining costs and many other advantages. In Al Gore's words: there is "genuine and realistic hope that we are finally putting ourselves on a path to solve the climate crisis." If we are, as he believes, at or beyond the proverbial tipping point in the transition to a solar-powered world, we might pause nevertheless to reflect on why it took so long to get here, and how best to sunset the privileges and immunities that have permitted the fossil-fuel industry to persist in what has become a war against humanity for the sake of making money that would have no useful purpose in a ruined world. We might also reflect on how best to ensure that concentrated and derelict power is never again allowed to compromise the safety and flourishing of humanity.[7]

I believe that we have the technical capability both to power the U.S. economy (and others) with a combination of steadily improving energy efficiency and solar energy, and to design hyperefficient solar-powered buildings and cities. The advantages of doing so include cleaner air, improved health, economic development, elimination of carbon emissions, less need for military engagement in contested parts of the world, no fossil-fuel money in American politics, improved balance of payments, and greater economic and ecological resilience. Disadvantages? Well, there really aren't any worth mentioning. I leave it to others to describe the many details of the transition, but we do not lack for well-thought-out possibilities, some already adopted by progressive governors, mayors, and NGO leaders, who are linking cities and regions together to reduce energy consumption and to adapt to unavoidable climate changes.[8]

But by the time we complete the transition to a solar-powered, carbon constrained world, we will be much further into the danger zone in which the effects of higher CO_2 levels will be evident along with results from various feedback loops, including the rapid release of methane from boreal forests and seabeds. The transition to renewable energy is only the first step in a long journey, and dangers will continue to multiply through the early phases, even as the levels of CO_2 and CO_{2e} (that is, all of the other heat-trapping gases measured in CO_2 equivalent units) fall, because of the mindboggling complexity and duration of the long emergency ahead. Throughout the transition we will need an unprecedented union of civic, scientific, economic, and political leadership as well as stable, competent governments. But again, other than the Catholic Church and the Chinese empire, we have few examples of institutional stability of the kind that we will need. To reach safe harbor requires us to solve the historical equivalent of a quadratic equation in which each of the parts must be handled in a particular order. For the sake of convenience, let's assume that the estimate by the Intergovernmental Panel on Climate Change (IPCC) is plausible and it will be approximately a thousand years before the Earth reaches a new equilibrium. In that time span, many things will occur that defy prediction. From the vantage point of one thousand years ago, for instance, our present world of seven billion people, smart phones, internet, air travel, and nuclear weapons would have been incomprehensible—and the next millennium will most likely be similarly incomprehensible to us. Looking forward, our progeny may have solved the major challenges of sustainability, peace, and equity and come through the bottleneck centuries intact and much improved. By contrast, they could be, as Jared Diamond and others warn, struggling on a barren planet graced with the ruins of our once global civilization.[9]

The difference between these futures depends on whether humankind can master four interlocked but distinctly different problems in ways that enable each to help solve the others. It is a systems crisis and must be solved in ways that each particular change fits into a larger whole so that the parts reinforce the stability, resilience, durability, and fairness of the larger system that includes the entire ecosphere: atmosphere, oceans, lands, forests, animals, and humankind. "We are faced not with separate crises," as Pope Francis writes, "but rather with one complex crisis which is both social and environmental."[10] It is a revolution in four parts.

The first and most important part, in Aldo Leopold's words, is a change in "our intellectual emphases, our loyalties, our affections, and our convictions," that is to say, a change of heart and our "habits of heart." Václav Havel writes similarly:

If we don't wish to find our world with twice its current population, half of it dying of hunger; if we don't wish to kill ourselves with ballistic missiles armed with atomic warheads or eliminate ourselves with bacteria specially cultivated for the purpose; if we don't wish to see some people go desperately hungry . . . If we don't wish to suffocate in the global greenhouse we are heating up for ourselves; if we don't wish to exhaust the non-renewable, mineral resources of this planet *we must—as humanity, as people, as conscious beings with spirit, mind, and a sense of responsibility—somehow come to our senses.*[11]

In other words, we need what theologian Cynthia Moe-Lobeda calls a "tectonic shift in moral consciousness" that grows into a force for systemic change over the long haul.[12] If that seems beyond our reach, remember that it has happened before. We no longer accept slavery; child labor; racial, ethnic, or gender discrimination; illiteracy; and other abhorrent behaviors once accepted as normal. As imperfectly realized as these may be, they are the standards by which we judge our behavior and measure our aspirations. Moral consciousness is changing similarly in matters pertaining to food, agriculture, consumption, energy use, the rights of future generations, and the role of humanity in the web of life. As we acquire a clearer moral vision, our awareness of possibilities also changes. What once seemed beyond our grasp can become the new normal.

Such a change will require a tectonic shift to a humbler, gentler, more thoughtful, and more ecologically competent human presence on Earth that begins by enough of us changing how we think and what we think about. The political ramifications are many. The conquistadors, militarists, terrorists of all kinds, captains of industry, connivers, dividers, high priests of hate, partisans, manipulators, snake-oil sellers, and demagogues have had their time. They have led us to an evolutionary cul-de-sac. If humankind has a better future it will be as a more "empathic civilization," one better balanced between our most competitive, hard-driving selves and our most harmonious, altruistic traits; one that embraces the yin-yang poles of behavior. It must be a change sufficiently global to bridge the chasms of ethnicity, gender, religion, nationality, and politics and deep enough to shift perceptions, behaviors, and values. The change must enable people to grow from a "having" to a "being" orientation to the world. It must deepen our appreciation, affiliation, and competence with the natural world, albeit a natural world undergoing accelerating changes.[13]

I do not think, however, that we can simply will ourselves to that empathic new world. The transition will result from social movements, activism, education, and political changes. But there is always an X-factor, an inexplicable process of metanoia, a word meaning "penitence; a reorientation of one's way of life; spiritual conversion." It is a change of inner sight. "I once was blind, but now I see" as the former slave trader John Newton wrote in the hymn "Amazing Grace." Metanoia is liberation from bondage—physical, mental, emotional—a total change of perspective. Viktor Frankl, for example, upon release from a Nazi concentration camp in 1945, wrote:

> I stopped, looked around, and up to the sky—and then I went down on my knees. At that moment there was very little I know of myself or of the world—I had but one sentence in mind—always the same: "I called to the Lord from my narrow prison and He answered me in the freedom of space."[14]

He then knew "the wonderful feeling that, after all he [man] has suffered, there is nothing he need fear any more—except his God."[15]

Metanoia implies a change of heart and perspective and also coming to see ourselves in a larger story as proposed by Thomas Berry. "The Earth is primary," he writes, "the human derivative." It came to us as a gift and it is to be passed on intact and whole to posterity. We exist in a sacred order of things, a chapter in "the Universe Story." But our role is unique: we are the species by which Earth becomes "conscious of itself in a special mode of reflective self-awareness." Our current predicament results from "a mode of consciousness that has established a radical discontinuity between the human and other modes of being and the bestowal of all rights on the humans." What Berry calls our "Great Work" is "the task of moving modern industrial civilization from its present devastating influence on the Earth to a more benign mode of presence." But to do so requires seeing ourselves in communion with nature in all of its forms, capable of making a response "that rises from the wild unconscious depths of the human soul."[16]

Building on Thomas Berry, Brian Swimme and Mary Evelyn Tucker place the story of humankind in a larger narrative as a journey from conqueror to participant in the "unfolding" cosmos. Perhaps, they write:

> human consciousness has a much larger significance within evolution than earlier philosophers could imagine. Could it be that our deeper destiny is to bring forth a new coherence within the planet as a whole, as the human community learns to align itself with the underlying dynamics of Earth's life?[17]

Swimme and Tucker note that "everything else in the universe seems to have a role . . . but "do we humans have such a role?" If so, it will be perceived in "Wonder[, which] is a gateway through which the universe floods in and takes up residence within us." It is not merely an emotion; it is an awakening to the paradox of how miniscule humans are on the cosmic stage but how important our role as creatures "with the capacity to feel comprehensive compassion."[18]

Wonder, however, can't be taught. There is no "Wonder 101" in the college catalog, and if there were, it would be a joke. But I think that we can make places and communities where wonder has a better chance of breaking through our psychological defenses. We can preserve the rituals, celebrations, and music that open us to wonder. We can preserve the sacred groves and sites where it has happened before. We can cultivate the awareness of possibilities that jar us out of our routines and open our sight. They will not likely be grand initiatives. A national program on metanoic transformation would never fly. No, this awareness will begin in smaller, almost invisible ways. So, how do we create the places and circumstances by which we come to our senses, causing a tectonic shift in moral consciousness?

Here we are helped by our natural, biophilic attraction to birds, bugs, trees, water, and land, and our "innate affinity for life and lifelike processes." It is the tug toward water, open spaces, plants, animals, and distinctive landscapes. It is the sense of place and connection that has inspired cave painters, poets, artists, nature writers, bird-watchers, zoo-goers, hikers, beachcombers, and small children through the ages. But it is perhaps only a tendency, and like others it can be corrupted, manipulated, or lost outright. Capitalists, advertisers, and increasingly sophisticated sellers of electronically contrived reality are working overtime to hijack our affections without ever quite admitting that to be their goal or at least a consequence of what they do.[19]

Like other human potentials, biophilia must be given room to grow and flourish. When it does we know that we heal faster, learn better, think more clearly, and are less stressed. Richard Louv, Wallace Nichols, Stephen Kellert, and others have compiled a large amount of convincing research on the good effects of exposure to nature through the biophilic design of buildings, landscapes, and communities. The story is the same in virtually all instances: we do better in the presence of sunlight, animals, growing plants, flowing water, and natural landscapes, and worse when such things are missing. When they are absent, the pathologies from "nature deficit disorder" multiply. The more technology in our lives, as Louv puts it, "the more nature we need." The exposure to nature also expands our sensory awareness and can increase life satisfaction. Again, this is not surprising. Our affinity for nature has been inscribed in us through thousands of years of evolution. Architect Bill Browning, sociologist Stephen Kellert, and urban designer Tim Beatley, among others, propose to reshape urban landscapes and building design to integrate nature throughout common daily experiences. But an expanded sense of affiliation for the natural world must be robust and tough enough to survive the hard times ahead.[20]

Natural systems, ecologies, and biodiversity will be severely stressed during the long centuries ahead. Many ecosystems will shrink. Some will die in place, dried out, burned over, flooded, or drowned under rising seas. Some species will migrate, but plants and animals are disappearing faster than scientists once predicted, and the loss of familiar species and places will cause mounting psychological trauma similar to what we would experience with the dying of old friends. We will all need care, counseling, and lots of hugs. Denying the reality of ecological death and sublimating our grief into even more severe pathologies such as violence or more zombie-like consumption is the worst thing we could do.[21]

Whatever we choose, however, time is not on our side. We must quickly learn to transform hardship into commitment, suffering into courage, and grief into action and get down to the work of restoration and repair. That requires changing the habits, strategies, and paradigms that once worked miracles but now threaten to be our undoing. We have intended to dominate nature but must now create a partnership. We have aimed for economic growth but must now make something different—durable, honest, and equitable patterns of livelihood. We have deployed technology without thought for the consequences but must now learn the discipline of precaution.

So far, we've been addressing symptoms, not underlying causes for reasons that recall George Orwell's words: "we all live by robbing Asiatic coolies. And those of us who are 'enlightened' all maintain that those coolies ought to be set free; but our standard of living, and hence our 'enlightenment' demands that the robbery shall continue. A humanitarian is always a hypocrite." We, too, have a good thing going that has been underwritten by structural evil we'd prefer not to notice, costs we'd prefer to pass to others, and changes that would be inconvenient to make.[22] So after all of the pious handwringing the time has come for a deeper level of honesty and what Pope Francis calls "a profound interior conversion."

No recent document has been more influential in describing that conversion and the paradoxes and challenges ahead than that written by Pope Francis in his encyclical *Laudato Si'*. It is a work of great scope and power that calls us to "a sense of deep communion with the rest of nature [and] "tenderness, compassion, and concern for our fellow human beings." The great power of the papal encyclical *Laudato Si'* lies in the union of concern for the health of the planet with concern for the poor. These are not, he writes, separate issues, but different sides of the same coin. He calls for an ecological-spiritual enlightenment that frees us from the domination of neo-liberal capitalism and its hardheartedness and short-sightedness that are destroying "our home." The great conceit of the modern mind is that we have risen above myth and superstition. But superstition seems to follow us like our shadow. We now measure and manipulate matter down to parts per billion and have become the most powerful force on the planet. Yet we are baffled by the rising tide of disorder all around us. Pope Francis's encyclical is a summons to pause, see, feel, and reflect on what we are doing before we undo our own humanity and the fabric of life itself.[23]

In a cynical world so saturated with the dogma of economic self-interest, *Laudato Si'* may read like so much tilting at windmills. But we might recall that our dreams of improvement once went deeper than merely the kind of technological advancement that strangles the larger imagination. Once we aimed for actual improvement in our organizations, institutions, economy, and governments. At the dawn of the European Enlightenment, some of the philosophers dreamed of a world run reasonably, by reasonable people, who gave good reasons for their actions—a world, in other words, without superstitions. In Thomas Paine's time (1737–1809), the most important superstitions had to do with arbitrary political and religious authority. We've done away with some of that, but other superstitions still multiply like rabbits. Here is a short list that includes only the most obvious:

- consumption makes our lives better;
- more stuff makes us happier;
- piling up mountains of stuff makes us a better nation;
- the Gross Domestic Product accurately measures our prosperity;
- our expenditure of $1 trillion on "defense" each year makes us safer; and
- owning guns makes a safer society.

So we shop as reflexively as our ancestors once worshipped. We degrade our air, waters, and lands in service to a deified market. We sacrifice the bodies and minds of our young to make the world safer for the rich. Ours is the ultimate age of superstition. Instead of unaccountable kings and priests, however, we are ruled by algorithms, corporations, billionaires, manipulators, and invisible technocrats with their mathematical models.

Laudato Si' is a call to replace economic superstitions and redirect our imagination toward transforming our motivations, purposes, and our circumstances. It is a call to create a new order of things: to imagine the human presence on Earth as a blessing and not as a wound, curse, or an affront to evolution. It challenges us to reconstitute a better and more durable civilization after the great flood of human-caused destruction has ebbed. It asks us to go beyond efficiency to sufficiency that leaves "as much and as good" to sustain the web of life in perpetuity. It requires us to raise future generations of planetary stewards with the heart and competence necessary to build a more durable, ecologically harmonious, beautiful, and just global civilization. *Laudato Si'* is a challenge to turn our ideals, sentiments, and hopes into a new reality that

- heals the land and waters;
- restores forests, habitats, and landscapes;
- grows our food sustainably;
- ends poverty and hunger;
- eliminates pollution and waste;
- builds in harmony with natural systems;
- shelters every human decently; and
- powers civilization by sunshine.[24]

Francis's vision is a union of new technology, old practices, integral design, and forbearance. It will also depend on a citizenry competent in the practical skills of conservation, farming, gardening, forestry, water purification, ecological restoration, land management, building, energy technologies, and urban and neighborhood planning. The goal is to create a new human presence on Earth rooted in an inclusive view of ourselves as citizens of an ecological community. The change requires the capacity to see the world holistically—a word linked not by accident to other words like whole, holy, hallowed, hale, hearty, and health. In its practical form, holistic management is the art and science of making the connections that bind complex ecological and human systems, acting skillfully at the right time in the right way, and with the patience to let nature do the rest at its own pace.[25]

Francis further enjoins us to create protected spaces, corridors, and sanctuaries for wildlife. It is a short step to a larger vision of rewilding the lands and seas of Earth as protective habitats as well as places of spiritual renewal. We have made a world in which we fit awkwardly at best. Among other liabilities of modern development many of us live in sensory deprivation zones called suburbs, and many more in badly designed cities overrun with cars. The shopping malls, sprawling development, necrotic urban expanse, industrialized landscapes, highways, and parking lots that we endure are noisy, ugly, polluted, barren places designed to accommodate machines, factories, and economic growth, not civilized humans. They are, as a result, inhospitable to a species that has lived 99 percent of its existence outdoors in forests, grasslands, and savannahs, and along lakes and rivers. "Our sublimated lives," George Monbiot writes, "oblige us to invent challenges to replace the horrors of which we have been deprived. We find ourselves hedged by the consequences of our nature."[26]

We need places of wilderness, as Wallace Stegner once wrote, by which we measure our human stature and find solace, mystery, and wonder. The preservation of large wild areas also constitutes a reservoir of species, a hedge against extinctions, and a baseline from which to measure the ecological health of developed lands. They are reminders of where we came from and so protect us from collective amnesia. Wild areas provide places by which we can see more clearly our proper role in the order of things and learn humility. They are refuges for increasingly rare possibilities for solitude, silence, reunion, and reflection. But why do we need reasons to preserve parts of the creation from human use? Aren't some places and some things sacred and so beyond our feeble, fickle, and time-bound rationalizations?[27]

It is clear that the quality and durability of our future depends on the creation of more interesting, beautiful, and sustainable places that integrate urban, rural, and wild. The standard for "development" should be what's good for children, songbirds, and human conviviality. Children need wildness most of all. Echoing Louv, Monbiot writes that "the collapse of children's engagement with nature has been even faster than the collapse of the natural world . . . the proportion of children regularly playing in wild places has fallen from over half to fewer than one in ten." As we've learned, the consequences of such a disconnect with nature include estrangement from each other and ailments such as asthma, obesity, heart and lung diseases, and autoimmune malfunctions. Isolation from nature also contributes to the feeling of disconnection, loneliness, and deep anomie that breeds nihilism.[28]

To such ends Dave Foreman and Michael Soulé propose continent-size restoration projects that include wildlife migration corridors connecting wilderness areas, parks, and cities. Edward O. Wilson proposes yet a larger vision by which humans would dedicate one-half of the surface of Earth to nature.[29] At whatever scale, the challenge is one to our foresight, convictions, and ecological imagination. Can we imagine buffalo once again as the dominant species on the Great Plains? Or rivers once again as fecund spawning grounds for wild salmon and sturgeon? Is our vision large enough to imagine the restoration of the Chesapeake Bay, the chestnut-dominated Eastern forests, or the vast schools of cod in the North Atlantic? Or at a global scale the restoration of the Aral Sea or Harappan forests of India? The necessary changes to do such things are not first and foremost matters of science, technology, or even money. They begin, rather, with a change of heart that expands our moral imagination and enlivens our compassion for all life.[30]

Many things stand in the way of the "profound interior conversion" and the foresight necessary to dedicate half of Earth to nature. The most important barrier is a belief that we can make end-runs around natural constraints. Accordingly, we must first disenthrall ourselves from the spell of total human domination of nature and its twin, the addiction to technology. "Man's attitude toward the world," in Václav Havel's words, "must be radically changed. We have to abandon the arrogant belief that the world is merely a puzzle to be solved, a machine with instructions for use waiting to be discovered, a body of information to be fed into a computer in the hope that sooner or later it will spit out a universal solution." But old, bad, and well-funded habits die hard. They persist by inertia and because they ask so little of us—only that we buy something newer, shinier, and even more digital. They offload our responsibilities onto others to discover and market "solutions" for us without requiring that we make an effort to understand the underlying causes of our problems and their relationship to other problems, or to consider other solutions, and even to change our behavior. Instead, we are offered only an even more fervent embrace with high-wire, indiscriminate, all-pervasive technologies that defy accountability and precaution. If things turn out badly, the consequences and risks inflicted by the unaccountable will fall on the innocent. Pope Francis calls this a "false or superficial ecology which bolsters complacency and a cheerful recklessness." Perhaps it is a modern version of blasphemy, where we act as our own gods. Or maybe it's just silliness on steroids, supported by a heavy dose of denial. But whether blasphemy or silliness, to those dying of thirst the effect is like a shimmering oasis in a desert mirage. We are always only a breakthrough or two away from one great triumph or another, but the breakthroughs sought are almost always the kind that will cause other perturbations, perplexities, and crises. This perpetual motion machine permits no discussion and no dissent; instead it demands only reflexive wonderment and "buy-in." It hides itself in a black jet of technical details that defy comprehension and demand faith—a faith that the things will work as specified, be managed responsibly in perpetuity, and will be just and beneficial, in some vague way. It is assumed that we will not read the fine print and that we have no capacity to direct the development of technology in ways that enhance our humanity and restore ecologies.[31]

In the shadows edging closer are human-made devices that are designed to be far more intelligent than we are. Again, no one has explained why on an overpopulated planet we need to conjure up a new form of quasimechanical beings intellectually far superior to ourselves that may not be "well disposed to us" and may develop consciousness of a kind we will not understand, and cannot control. If they do prove to be uncontrollable, what would happen to us, mere humans? Before even crossing that line we would do well to ask whether this is where we want to go. Who was given the right to play god over humans' future? And for what reason? Was the purpose all along to create robots that would be cheaper and more obediently lethal than flesh and blood soldiers? And if it turns out badly, as it may, who will be held liable? The self-displacement of humans by intelligent, perhaps sentient machines that might harm us may, in the end, be among the most droolingly stupid things ever attempted. But even if robots are constrained only to do our work and improbably follow Isaac Asimov's rules for treating their masters kindly and carefully, what will we then do with the millions of people permanently unemployed? With what money would they buy whatever it is that robots would make, grow, or service?[32]

The first challenge requires a change in our loyalties and affections, that is, a change of heart. The second requires a corresponding change in how we think and what we think about. But thinking clearly and cogently is among the hardest tasks that humans ever attempt. Our mental shortcomings are many. We see the shadows on the walls of the cave, but are baffled about their meaning. We speak and perhaps think in clichés. We are as vulnerable to intellectual fads as we are to the bizarre whims of high fashion. We are gullible and often ignorant of the most basic things. Many of our fellow citizens, for example, believe in alien abduction, emphatically doubt the physics that govern the climate, and think the science of evolution is a fraud. Even the more thoughtful among us are blinded by blizzards of information and bombarded with artful lies, and so find it hard to separate truth from falsehood, the trivial from the important, or wisdom from folly. We are a people ripe for the plucking standing in a crowd of pluckers.[33]

The reasons for our shortcomings are due in part to our origins. We evolved in an ecologically more complex but socially less complicated world, in which we told stories to each other around tribal campfires and village greens. Our great explainers were storytellers, shamans, witch doctors, pathological control freaks, serial sadists posing as patriots, and some men who explained things to their womenfolk. But the damage was lessened somewhat by a few saints, true philosophers, and people with good sense and good hearts. In any case the ruin done was infinitesimally small compared to the havoc that one deranged national leader, jihadist, corporate CEO, or psychopath can cause now.

We live in a tightly interconnected but also highly fractured nuclear-armed world in which our very survival depends on our learning to overcome our various parochialisms and divisions. We may never learn to see each other as brothers and sisters, but we have to learn to get along. That is not so much a breakthrough as a long process by which we learn empathy and acquire the art and science of systems thinking—by which I mean the quality of mind that discerns the "patterns that connect," in Gregory Bateson's words. The ability to see our connectedness in larger systems is inherent in all religions, the root word for which means bound together. Like it or not, the fact is that we are kin to all that ever lived and all that ever will live—one link in the great chain of being. We may not appreciate all of our kinfolk, but their pictures are in the family scrapbook alongside our own. We are a small thread in the fabric of evolving life on Earth. In the presence of such vastness and mystery the only appropriate attitudes are those of wonder, gratitude, and lots of humility. But this is not what modern education aims to cause or cultivate.[34]

Higher education in the Western world was conceived mostly as a means to propagate Christianity and somewhat later as a means to enlarge the human estate and solve the problem of scarcity. The purpose of education presently is not to foster wonder or gratitude or ecological competence but rather to equip young people for jobs and careers in an economy designed to expand without limits. As Thomas Berry puts it:

> The university prepares students for their role in extending human dominion over the natural world, not for intimate presence to the natural world. Use of this power in a deleterious manner has devastated the planet . . . so awesome is the devastation we are bringing

about that we can only conclude that we are caught in a severe cultural disorientation, a disorientation that is sustained intellectually by the university, economically by the corporation, legally by the Constitution, spiritually by religious institutions.[35]

The purpose was domination of nature up to the heavens and down to the nanoparticle, including everything in between. Living in harmony with nature was not a part of it. The machinery of education from Francis Bacon's time to our own was designed to generate useful knowledge aimed, as noted earlier, at mastery, not of ourselves so much as "the affecting of all things possible." But as C. S. Lewis once commented, that mission was born in irony:

> At the moment, then, of Man's victory over Nature, we find the whole human race subjected to some individual men, and those individuals subjected to that in themselves which is purely "natural"—to their irrational impulses . . . Man's conquest of Nature turns out, in the moment of its consummation, to be Nature's conquest of Man.[36]

The result of the "manipulation of limited knowledge, brute force, and an overwhelming arrogance" has put civilization itself in jeopardy.[37]

The modern university has come to resemble a Maginot Line with separate fortresses surrounded by moats and minefields. Against nearly impregnable fortifications, direct assaults are almost always futile, especially when organized and led by the nontenured. When it advances, knowledge does so indirectly by flanking maneuvers and what Thomas Kuhn describes as paradigm changes, but even so, the defense has the advantage. The more adept and aggressive disciplinary fiefdoms vigorously exclude threatening paradigm changes by gerrymandering the boundaries and patrolling the borders to prevent either defections or intrusions. In fact, all disciplines in higher education endeavor to maintain a monopoly of terms, theories, and agendas, and befuddlements for the un-credentialed. It is called rigor, but is often hard to distinguish from rigor mortis. For all its needless complexity, Rube Goldberg might have been thought the architect of the byzantine machinery of knowledge "production" and transfer,

THE LONG REVOLUTION

but alas it was done as absent-mindedly as the British once acquired an empire. Nearly everywhere the results are the same. Our academic efforts are generally centripetal, focused (even in this day of "interdisciplinarity") on the problems narrowly defined by discipline and subdiscipline. And the discussions among the professoriate, with some notable exceptions, leave aside messy, big questions about the fate of civilization and human survival that are beyond this year's departmental budget and the pressing concerns of parking permits.[38]

It is not surprising, then, that higher education has lost its way and the reasons are many. It is too expensive and too oriented to careers. The system is often demoralizing to students and faculty alike, populated by a growing number of underpaid and exploited adjunct faculty and administered in its upper reaches by educational barons paid princely salaries with lavish perks. In the larger and more successful universities, the ancient purposes of learning and so forth have become adornments to the sports programs that rake in millions of dollars while budgets for the philosophy department are slashed. Those who believe that markets are the answer, whatever the question, might propose that philosophers field their own teams that would play in stadiums suitably named for Immanuel Kant, René Descartes, Socrates, or maybe even Milton Friedman, paid for by sales of books on the meaning of life and other deep subjects. The possibilities are many. The pecuniary rewards, alas, are not so promising. But I digress.[39]

So, what is the "value added" of four years of higher education? What do the higher educators do that justifies an average of around thirty thousand dollars of student debt at graduation? What opportunities do we open to our students, not just for jobs but to do good work? Mostly, I fear, it is more of the same. The choice of "training persons for temporary survival in the declining Cenozoic Era or . . . for the emerging Ecozoic," as Thomas Berry puts it, is really no choice for those current vendors of career guidance. Fame, fortune, and presumably future largess to one's alma mater still depend overwhelmingly on the Cenozoic-era careers connected to the financial "industry" and to the extractive economy. In our learned conversations we seldom ask what the transitional generations will need to know in the long emergency ahead or how they might help in the transition to the Ecozoic era. Will they need to know more about information technologies and contribute to the flood of information? Should they aim to become masters in the art of communication but about what great ideas?, or gene splicing with what ecological foundation?, or science, technology, engineering, and mathematics (STEM) untempered by the humanities? What should they know in order to make wise choices about technology? Or will they need other skills that equip them to build just and prosperous economies that do no collateral harm, to grow a generous heart, to serve their neighbors, or simply to live well in their place? Is the goal to create an endless supply of hypersmart but footloose vandals to prey on the world's last forests, fossil fuel reserves, minerals, fish, genetic materials, and public gullibility? Or is it to educate and nurture healers, fighters for justice, ecologically savvy farmers and foresters, artists and musicians, thinkers, philosophers, builders of cultural and social bridges, and visionaries who can see what isn't but should be and make it happen?[40]

Beyond the primary obligation to serve students, what is the larger role of colleges and universities as "anchor institutions" in their own communities? Many communities have been devastated by deindustrialization, pollution, urban sprawl, and a flood of drugs. Climate change will make these worse and bring massive new problems. In this light what is the role for institutions of higher education? Michael Crow and William Dabars call the "New American University" one that "takes major responsibility for the economic, social, and cultural vitality of its community." Can colleges and universities redirect their buying and investments to local and regional renewal? Can they become carbon neutral and lead others to the same goal? Can they help to restore regional agriculture, food systems, and businesses? Might they invest in restorative economic development? In short, can they operate facilities and manage their financial affairs in ways that do no harm to the world their students will inherit? And can they make these a part of their curriculum so that the young become competent in making high ideals practical and real?[41]

In response to such questions, a transformation of higher education is now under way. Construction of the Adam Joseph Lewis Center at Oberlin College in 2000 was a major catalyst for the green building revolution on college and university campuses. The building is powered entirely by solar energy, even in a part of the country where sunshine is mostly a theory. Organizations such as Second Nature founded by Tony Cortese have secured commitments by hundreds of college and university presidents to become carbon neutral. Julian Keniry and Kevin Coyle at the National Wildlife Federation and others are working to reshape the curriculum in higher education. Mark Orlowski's "billion dollar challenge" aims to shift university endowments and investments away from fossil fuels and toward renewable energy. And the American Association for Sustainability in Higher Education aims to make sustainability the guiding principle of colleges and universities.[42]

In short, there is a great deal of activity and lots of good news in the higher reaches of education, but the tide has not yet turned. The dominant educational paradigm is still rooted in the industrial model that aims to extend humans' mastery over nature. The conversations in the academy are still more notable for what they do not include than for their breadth. There is still little mention of the core assumptions of education, or those of underlying science, or those driving technology, or about alternative pedagogies. Such things get too close to the heart of the problem for the comfort of the comfortable. Most presidents, chancellors, deans, administrators, and trustees calmly regard the transition ahead as a fine-tuning of more of the same. Even at this late hour few see the challenges of sustainability as reasons to make fundamental changes in curriculum, pedagogy, structure, operations, and finance. The system of higher education still rewards careerists and courtiers all too well. So, for all of the new science facilities, bio-tech labs, sports arenas, and glitzy business-school buildings, we do not know whether formal education will, on balance, be a positive factor in the long journey ahead, or neutral, or even negative.[43]

For all of these reasons, to find real educational innovation you must go to the periphery, both inside and outside the academy. Changes are mostly driven from the margins from small, under-funded institutions like Schumacher College in Devon, England; the Aldo Leopold Center in Baraboo, Wisconsin; the Center for Ecoliteracy in Berkeley, California; the Center for Environmental and Sustainability Education at Florida Gulf Coast University; and hundreds of other small nonprofit organizations around the world. The emerging pedagogy also comes from thinkers outside the mainstream like Satish Kumar, Chet Bowers, Nicholas Maxwell, and Gregory Cajete, who blend ancient wisdom with modern science, emphasizing connectedness at all levels.

Changes to our hearts and minds are internal to us. At the societal scale, they are slow tectonic shifts in thought and behavior, that is, until they spark social movements and revolutionary behavioral and political change. They proceed heart to heart and mind to mind changing what we care about, what we pay attention to, how we speak, what we talk about, and how we act. They are the broad currents that flow below the surface of history. They can be progressive as, say, the Enlightenment that spread across Western Europe in the eighteenth century or as regressive as the troubling eruptions of xenophobia, religious fanaticism, and racial violence that punctuate human history.

The third change necessary to the transition is public, not private: it has to do with power, politics, public affairs, government, and issues of war and peace. But any discussion of politics and government must begin with the caveat that there is no "altogether workable solution to a not altogether solvable problem." That is, the chief obstacle to good governance and to the political process is the perennial problem of concentrated power that in Frederick Douglass's words "concedes nothing without demand. It never did and it never will." Governments and the conduct of the public business are measured against the challenges of the time, the particularities of culture, and the standards of effectiveness and fairness that vary in different times and places. But the standards of administrative competence, transparency, justice, and foresight against which all governments must be judged never change.[44]

The long emergency ahead will require these qualities of good governance and that our civic life, public business, and politics be harmonized with the way the Earth works as a biophysical system. The great irony of the libertarian spasm of our time is that the longer it takes for governments and political systems to align with the way the Earth works as a biophysical system, the larger and more intrusive they will have to become. The libertarian—and also Marxian—fantasy of governments withering away is likely to lead to the largest expansion of centralized authority ever. And it is a good bet that is where we are headed. Although they can do many things, markets are no substitute for competent governments. If you doubt that, try calling your nearest Fortune 500 company for help as a Sandy- or Katrina-scale storm is breaking over your head or prolonged heat and drought is blistering your land.

Under conditions of prolonged ecological duress, competent governments are the only possible bulwark against societal chaos and collapse. Without significant changes in how we conduct the public business, however, they are likely to become ever more intrusive and authoritarian. It might have been otherwise had we come together and acted earlier while there was still time enough to think it through. But that is so much water under the bridge. The result of the decades of fossil-fuel-funded denial is that governments will necessarily have to reckon with a long list of issues that were once problems but have now become protracted global crises. For all governments, the conditions of the long emergency pose four very large challenges.

The first is that of harmonizing increasingly complex and dispersed government functions with the larger ecological systems in which they're embedded. Ecosystems work as integrated wholes, with the parts reinforcing the health and resilience of the entire system as it evolves over time. Governments, by contrast, are composed of fragmented bureaucracies, legislatures, agencies, and courts with competing interests and often working at cross-purposes as they compete for resources. The long emergency will require governments to act harmoniously within the larger patterns and time scales of natural systems.

If that isn't daunting enough, government officials appointed and elected must be ecologically literate enough to understand how the Earth works as a biophysical system and how that knowledge affects what they do as administrators, legislators, and judges. In recent years, unfortunately, the slightest quiver of synaptic activity among public servants in those parts of the brain having to do with ecological cause and societal effect has been regarded as a great embarrassment. Going forward, this condition must be remedied by ecologically literate and civically inclined voters who understand why elections matter for their lives and livelihood and for their children's prospects on a small planet.[45]

Further, laws and policies will require coordination between agencies, for example, between the departments charged with management of the economy and those charged with cleaning up the resulting messes. The National Environmental Policy Act of 1970 was a first step toward ecologically coherent government, but it has been rendered ineffective due to a lack of appreciation and support by those with strong emotional and financial ties to the extractive economy.[46]

The second challenge to government is to ensure social stability and order for long time spans and under circumstances that will grow steadily more difficult. Climate destabilization will cause famine, droughts, large storms, larger forest fires, millions of ecological refugees, and flooded coastal cities, along with the usual problems of economic stability, war and peace, terrorist acts, and periodic manmade disasters. The U.S. Constitution is less than two hundred and fifty years old and has already survived longer than James Madison, for one, thought possible. The British Constitution has been evolving since the Magna Carta in 1215. The Catholic Church is roughly 1,600 years in the making and still ministers to the spiritual lives of 1.2 billion members. The Chinese empire lasted for several millennia depending on where one starts counting. These examples should give us encouragement that it is possible to create stable, if not perfect, systems of government resilient enough to prevail through increasingly turbulent times—even if the longevity of the challenge and the sheer scale and severity of the situation are unprecedented.[47]

A third challenge to government concerns the rights of future generations and eventually to other species. Our political history is a record of painfully slow progress in the area of extending rights to previously excluded groups. It requires no great leap of logic or philosophy to take the next step and admit future generations into the great civic community as bearers of rights that cannot be infringed without due process. Our descendants will be as human as we are. Like us, they will have basic physical requirements for clean water, food, shelter, employment, and sociability. Like us, they will prefer to lead dignified lives rather than exist as servile objects. They, too, will express themselves in art, music, poetry, dance, film, and literature. But unlike us they will have no power to forestall the reality of a deteriorating climate and a biologically impoverished planet, as well as the suffering that will come with it. They will have heard stories of a time long ago called the Holocene, but their lives will be lived on a different kind of planet. It is worth noting that, barring a catastrophic population collapse, future generations will outnumber those presently living by orders of magnitude. In an intergenerational democracy, they would presumably outvote us on matters that threatened the ecological conditions necessary to protect their rights to life, liberty, happiness. But our democracy does not grant them such rights or even standing in courts. Unless we act to reduce the scale and duration of the calamity and protect their persons and interests, their lives, indeed, could be nasty, brutish, and short. The choice is ours and ours alone. To act on their behalf does not require an extension of the basic logic of rights to persons of other nationalities, ethnicities, gender, ages, or sexual orientation. But it does require that we extend our affections, sympathies, and protections in order to forestall unnecessary future suffering. The reasons to do so are not entirely altruistic. The fact is that the suffering from climate destabilization has already begun. In 2013 alone, the United States experienced $125 billion in damages from climate-change-driven weather disasters. And that is only the start. The longer we delay effective climate action, the worse it will get, for ourselves and for succeeding generations.[48]

It is a different case when we consider extending rights in some form to other species. They are unlike us, but not entirely. They cannot reciprocate, but only as far as we know given our present knowledge. With a few well-studied exceptions they cannot communicate with language, but do so in other ways. Even so, the more we learn about other species, including plants, the more it seems that intelligence or something like it is widely diffused in nature. The closer we observe, the more often we find behaviors that resemble our own. Is it possible that intelligence pervades the entire system? Although it is hard to know what to make of James Lovelock and Lynn Margulis's Gaia hypothesis, it is clear that enough scientific evidence exists to suggest that the Earth has some of the characteristics common to all living systems—a possibility discordant with our reductionist and anthropocentric tendencies.[49]

Following this line of thought, is it possible that our moral progress requires extending consideration to members of all species, as Charles Darwin thought? Is it possible, for example, that our industrialized animal gulags and mechanized fishing fleets that feed us so cheaply are an affront to a larger moral order? Could it be that something like an apology or even restitution is due? And is it possible that we live in a larger ecology of inescapable irony and paradox by which the race is not to the swift, nor the battle to the strong? The Taoists hinted as much, thinking water the strongest of all elements.

For the practically minded, the case for extending consideration to other species can also be made on more familiar grounds on which the rights of all species to exist are combined with enlightened self-interest. That is, in addition to the case for the intrinsic rights of other species regardless of their usefulness to humans, we can look at the benefits to us that these other species provide and decide that these "ecosystem services" provide a strong utilitarian reason for their preservation. In a well-known and controversial study, Robert Costanza and his colleagues documented $33 trillion in "services" that natural systems provide for free. But this number excludes much that nature does for us that we don't comprehend, as well as those things that are priceless. Among these one could include the benefits of companionship and spiritual nourishment that other forms of life provide. It seems to me that the loneliness and despair that would result from living in a world without other life forms would drive us over the edge, if nothing else had by then. But no economist has yet run the numbers.[50]

The fourth challenge to governments has to do with public distrust. Given the growing importance of government to a decent human future, we must outgrow the simplistic and debilitating idea that government is always and necessarily a problem. Governments certainly can cause trouble and often throughout history they have. But the same is true of errant political parties, venal corporations, secret societies, virtually all religions at some time in their history, and groups of chest-thumping, gun-toting, hyperpatriots. Human organizations are imperfect things. But in whatever form, governments, in particular, are indispensable, even if they must be held accountable, and be democratic, transparent, well-organized, and properly led—traits we associate with the high points of our political history.

Confronting the long emergency, we have no alternative but to greatly improve the efficiency, accountability, foresight, and leadership of governments at all levels. To do so, we will need a more effective, resilient, and thoroughly democratic vision of the state, as well as a more inclusive political vocabulary. Philosopher Peter Brown, for example, proposes that we return to the idea that government is like a "fiduciary trust in which the trustee has a duty to preserve and enhance the assets of the trust with a view to the good of the beneficiaries: the citizens . . . [charged to] preserve and enhance the 'property' of all citizens." The model further requires that governments, as Trustees, act impartially and protect human rights, prohibit waste, address "crises of natural resources and biostability," respect the "virtues of commerce," and "make it explicit that our obligations extend through time." The trustee concept of government is an old idea that bridges conservative and liberal views of government. It is liberal in the current sense of the word: that is, it holds that governments ought to protect the health and well-being of all citizens, particularly the most vulnerable. It is also conservative in that it protects customs, religions, laws, and institutions that have enduring value. It is rooted in the precedents extending back to the Roman *Institutes of Justinian* (535 CE), the Magna Carta (1215), and hundreds of court decisions in the United States and elsewhere. It is embedded in some state constitutions and a growing body of international law. It entails the delegation of authority from the people to government to manage, protect, and preserve public assets for the public, not private, benefit. In legal scholar Mary Woods's view, the ecological trust includes "natural wealth that must sustain all foreseeable future generations of humanity." Applied to natural systems, the trust doctrine includes all aspects essential to healthy ecosystems and can be said to be holistic. Unfortunately, courts have been inattentive to the science of ecology, persistently dividing ecological systems into various asset categories. That is, legal scholars, lawyers, and judges, with a few notable exceptions, have been conspicuously absent, and sometimes retrograde, on issues pertaining to the habitability of the planet on which their own offices and courts are situated.[51]

The public trust doctrine would further advance the role of government in managing change. For example, the federal government has historically been the leading entrepreneurial agent in society, funding "the riskiest research, whether applied or basic, but it has indeed often been the source of the most radical, path-breaking types of innovation." Canals, railroads, highways, bridges, air travel, and the internet all depended on government investment at early stages when they were deemed too risky for private investors. The iPod, iPhone, iPad, and similar products, for example, depended on early research and funding from the U.S. government. The government (that is, the public), however, has been neither appropriately compensated, nor received even a whisper of gratitude, from those who have profited so handsomely. Adding insult to injury, the parent company now pays the merest pittance in taxes. The transition ahead will require similar federal support for research and deployment of break-through renewable energy technologies, a modern smart electric grid, materials recycling, cradle-to-cradle materials systems, water desalinization plants, and sustainable agriculture. Economist Mariana Mazzucato proposes that the public, in return for its investment, should be able to extract a royalty payable into a "national innovation fund" to support other early stage development and deployment. Most investors do expect a rate of return. Why should it be different, she argues, for the public?[52]

Larger and more difficult challenges are on the horizon. In the years and centuries ahead a still-growing population must be fed, watered, sheltered, transported, and provisioned from a resource base reduced by drought-afflicted and thinner topsoils, declining aquifers, depleted resource stocks, burned-out forests, disrupted ecologies, and flooded coasts. With our world population expected to peak at perhaps eleven billion, we will need to find a better criterion than "ability to pay" to decide who can have which goods and services. Stan Cox of the Land Institute in Kansas puts the problem this way: "If we do make a serious effort to corral the national and world economies within ecologically supportable boundaries, the method we choose for divvying up the resources that humanity can afford to consume must be a method we can all live with." A mad scramble for what's left is a prescription for catastrophe in which only the wealthy will win, and only for a moment. In Cox's view, "it is not a matter of whether we ration but rather how." In other words, a "people of plenty" must become a more restrained and smarter people befitting ecological scarcity. Fair distribution in the circumstances of the long emergency will require us to rethink what constitutes basic human needs, good work, and how to hold things together in an age of consequences.[53]

Rationing, whether by price or regulation, speaks to a larger problem of property and ownership. Throughout much of human history, property did not exist in the modern sense. Instead land, waters, wildlife, and much of culture were held as common property owned by no single person, thereby affording widespread rights of usufruct. The rise of what Andro Linklater calls the "private property society" in the sixteenth century ushered in a "monstrous method of owning the earth." But a system of private ownership, it turns out, works only as long as public interests are given equal weight, much as once specified by John Locke, Adam Smith, and James Madison. As history would have it, that caveat was conveniently ignored. Beginning first in England in the sixteenth century, commonly held lands and forests were appropriated by wealthy owners who displaced the customary users. Legal doctrine dutifully followed. In his famous 1968 journal article "The Tragedy of the Commons," Garrett Hardin made this aberrant history into an argument that the fate of all common property regimes was ruin. The reason lay in his assumption of a world populated by the standard-issue economic man who would indeed ruin virtually any commonly owned property available. The solution he proposed was "mutual coercion, mutually agreed on," that is to say, either private ownership or government control. The theory, however, differs from the actual history of common property management. Political scientist Elinor Ostrom, for one, uncovered a different history in which commoners, over time, evolved rules for the management of common property and thereby managed resources sustainably over long periods of time. She concluded that capitalism amplified by hyperindividualism is the force that ruins the commons.[54]

The question is whether common-property management can be resuscitated as "the basis for a larger, macro-solution without some new legal and policy architecture that can recognize and support the skillful nesting of different types of authority and control at different levels of governance." Legal scholars Burns Weston and David Bollier argue that the state on its own is not up to the task of saving the planet. We must, therefore, think and organize ourselves more creatively. The ancient traditions of common-property management still exist, for example, in community water systems (*acequias*) in New Mexico and seed-sharing commons managed by *dalit* women in the Andhra Pradesh region of India. In these and many other instances the commons approach to property rights "is a structural commitment to managing resources in ways that minimize social inequities and ecological harm," but it is alien, they write, to the "outdated twentieth-century mind schooled in traditional, top-down bureaucratic control." Management of our common property of forests, lands, game, or water does not fit the model of rational control and linear causality because "they are subject to too many incalculable and qualitative variables in ways that blithely transgress established political boundaries." But these methods worked effectively over centuries in many places to protect lives, livelihoods, and ecologies alike. In recent U.S. history, the Alaska Permanent Fund, in place since 1976, distributes royalties from oil drilling on state lands to residents. Larger possibilities to expand the public trust doctrine and common-property management also exist.[55]

Peter Barnes, the founder of Working Assets, proposes the creation of an American Commons Fund funded by the sale of emission permits for carbon dioxide pollution and eventually from dividends from publicly traded corporate shares. His goal is not to change human nature, but rather to "define new property rights and boundaries" similar to those we commonly accept, like height limits on buildings, speed limits on highways, and Plimsoll lines on the hulls of ships to prevent overloading. He further proposes to create a holding company, Earth Holdings, Inc., that would own the sky, the oceans, the aquifers, DNA, and other common assets in trust for humankind, funded by taking money from "overusers of Earth's assets."[56]

Barnes, Weston, Bollier, and others argue that the implementation of a "new ecological governance system is tantamount to a fight for human survival." That struggle begins with the realization that nature is not simply a commodity to be owned in order to make profit as "we once bought and sold human beings at auction." How did the few seize control of so much? By what means can humankind take back its ancient prerogatives and responsibilities for tending to their lands, forests, wildlife, waters, and genetic materials? Can common-property regimes be scaled up to become an important part of a new system of governance for the Anthropocene?[57]

If the answer is in doubt, there are other possibilities. Ostrom argues, for example, that top-down, centralized approaches to climate change in particular are inherently weak because of what economists call the "free-rider" problem, in which some receive the benefits of group action without actually contributing to the outcome. It is a problem that afflicts all committees, civic organizations, and international organizations where the inactive, lazy, incompetent, and disinclined nonetheless reap all of the rewards of group action. Ostrom argues that it is best solved in the case of climate change by "polycentric approaches" at multiple scales and levels below that of a central authority. The examples are many and instructive. The U.S. Green Building Council, for one, has markedly improved building practices and energy efficiency without a government mandate or public sanctions. Rather the council successfully developed, propagated, and supervised an evolving building-rating system that spread by peer pressure, superior results, and the power of networks that included architects, engineers, builders, city officials, the construction and building materials industries, and energy technology companies. Other examples include nongovernmental organizations that enforce standards, such as the Forest Stewardship Council, or that disclose otherwise private data such as the Carbon Disclosure Project and the Carbon Tracker Initiative.[58]

Showing the leverage of near-monopoly power, WalMart, for one, required its many thousands of suppliers to reduce packaging, energy use, and CO_2 emissions. Whatever one may think of the company, it has on its own reduced its carbon footprint and environmental impacts, causing ripple effects throughout the economy. Numerous similar examples of "private environmental governance" or "polycentric climate governance" show "that public action is not the only way to achieve public ends" and demonstrate "new ways to think about collective action problems . . . and sources of pressure on firm behavior other than government regulation." Even so, there is and always will be a need for a flexible but robust centralized authority to set and enforce the rules in which these polycentric possibilities can coexist with other ways to do the public business.[59]

As I write these words, it is March 2016 and our common democracy is in disrepair. In some places voting is being made more arbitrary and difficult. Anonymous money floods into campaign coffers to work its mischief. And the U.S. political system is in yet another of its quadrennial spasms of political silliness. The Republican candidates dance awkwardly like puppets to the tunes played by a few billionaires who have been licensed by five members of the Supreme Court to purchase the election: in return for vast sums of secret money, their candidates will agree to testify for an alternative reality in which the laws of physics and chemistry do not perturb the atmosphere and those of biological evolution do not trouble the presumed archetypal human created by a fairly dogmatic God. Further, the candidates must subscribe to the view that the poverty of the poor is self-induced and has nothing to do with economic or tax policies that enrich the rich and impoverish the public sector. Such candidates for the highest office in the United States must also be able ringmasters in a circus of distraction. The public is not to be informed, educated, or summoned to higher purpose, including its own survival or that of civilization. And the agencies of democratic government, accordingly, must be corrupted, undermined, and abused. In this way, the theft continues and, sadly, many of those being robbed cheer it on.

In his final book, economist John Kenneth Galbraith wrote: "Human progress [is] dominated by unimaginable cruelty and death . . . Mass slaughter has become the ultimate civilized achievement . . . War remains the decisive human failure." The consequences of giving "a privileged position to the development of weapons and the threat and reality of war" would not have surprised James Madison, except for its sheer scale. Madison feared standing armies that "with an overgrown Executive will not long be safe companions to liberty. The means of defense agst. [*sic*] foreign danger, have been always the instruments of tyranny at home." A few years later Madison elaborated:

> Of all the enemies to public liberty war is, perhaps, the most to be dreaded, because it comprises and develops the germ of every other. War is the parent of armies; from these proceed debts and taxes; and armies, and debts, and taxes are the known instruments for bringing the many under the domination of the few. In war, too, the discre-

tionary power of the Executive is extended; its influence in dealing out offices, honors, and emoluments is multiplied; and all the means of seducing the minds, are added to those of subduing the force of the people. The same malignant aspect in republicanism may be traced in the inequality of fortunes, and the opportunities of fraud, growing out of a state of war and in the degeneracy of manners and morals engendered by both. No nation could preserve its freedom in the midst of continual warfare.[60]

More than two centuries later, the American security state is in full stride even after the demise of its arch enemy, the Soviet Union. It spends more than all other nations combined on defense. The Pentagon's annual budget along with allocations for wars is perhaps $1.3 trillion, for "defense" and perhaps as much again to snoop and spy, but no one knows either number for sure and few in Congress ever ask for an exact figure or inquire "how much is enough." The U.S. Department of Defense is said to maintain over one thousand bases around the world. It patrols the seven seas and is prepared to fight multiple wars simultaneously, but against whom is unclear. It exists to fight a long list of dependably loathsome enemies—some conjured, some real, some domestic, some foreign, some with legitimate grievances against us, and some who pose real threats to us without good cause. The war system, including weapons suppliers, consultants, contractors, the Pentagon, and members of Congress, depends greatly on permanent threats that are real or, in various ways, manufactured. It is administered by a double government: the visible part consists of elected officials that appear on television, whereas the invisible, and more powerful, part consists of several hundred members of the "national security culture"—careerists, political appointees, academics, think-tankers, and military officers over "which even the President now exercises little substantive control." It is this invisible part of the state that sets the agenda and strategy. The national security state was born in the National Security Act of 1947, which unified the different branches of the military and created the Joint Chiefs of Staff, the Central Intelligence Agency, the National Security Council, and the National Security Agency. In total, forty-six federal departments and agencies are engaged in national security work, cyber operations, weapons development, and war-related fighting. The system is held together in political scientist Michael Glennon's words by "loyalty, collective responsibility, and—most important—secrecy," to which it is addicted. It does not tolerate dissidents like Edward Snowden, whistleblowers of any kind, or contrary worldviews from the likes of Martin Luther King, who opposed the Vietnam War, and more recently international legal scholar Richard Falk. Its participants prefer stealth and force over diplomacy and peace-making. This mindset of militarism permeates the gun-toting elements that lurk in the shadows of American democracy; note the celebration of violence and the kind of knee-jerk, flag-waving, mindless patriotism of the sort that Samuel Johnson called the "last refuge of scoundrels." The entire supply chain of violence includes weapons manufacturers, subcontractors, and consultants; the domestic gun and ammunition manufacturers; and even the bereavement industry, which provides the flowered caskets in which we bury the bodies of our 34,000 or more gun violence victims every year. It reaches into every Congressional district, a fact much on the minds of the members of Congress. And it has acquired the power that James Madison most feared:

> The power to kill and arrest and jail, the power to see and hear and read peoples' every word and action, the power to instill fear and suspicion, the power to quash investigations and quell speech, the power to shape public debate or to curtail it, and the power to hide its deeds and evade its weak-kneed overseers . . . in short the power of *irreversibility*.[61]

If one asks why the United States has sunk to the bottom of most social indicators for well-being of its people, or why our infrastructure is rusting away, or why 22 percent of our children live in poverty, or why so many of our young black and Latino men are in prison, or why we lead the world in homicides and mass killings, or why many of our once finest cities would embarrass residents of third-world nations, or above all why we have been paralyzed in the face of the self-generated threat of climate instability, a major reason is the hold that militarism and violence have on our hearts, minds, pocketbooks, and the national checkbook. We have become a warlike people living in what increasingly resembles a garrison state, continually surveilled and taxed to support a vast, unaccountable, military-corporate complex that aims to dominate the world. It is a fool's errand, as history has repeatedly shown. Paradoxically, too, Americans are the most feared and possibly the most fearful people on Earth. As former Army colonel and professor of history Andrew Bacevich states, the time has come to "consider dismantling an apparatus that demonstrably serves no [good] purpose."[62]

The first priority to that end is to abolish nuclear weapons as proposed by former secretary of state Henry Kissinger, former secretary of defense William Perry, Senator Sam Nunn, Stanford physicist Sidney Drell, former secretary of state George Schultz, and General Lee Butler who once presided over the U.S. nuclear forces.[63] The nuclear weapons that they propose to abolish, however, are the logical development of the war system—and this vast, complex, hugely expensive, life-denying, and dangerous machine has been designed to perpetuate itself at public expense and enrich its progenitors. To an extent difficult to ascertain, it generates the threats that it purports to defend us against, which in turn begets yet more threats, violence, and profits. The dynamics of perpetual conflict have been known since Thucydides studied the Peloponnesian Wars. But in a world of spears, bows and arrows, and swords, war posed no global threat, however imbecilic and futile. In a global world armed with nuclear weapons, however, nonviolent conflict resolution is the only practical alternative to eventual Armaggedon. Peace is no longer a romantic dream but the only practical option we have left.[64]

Few have thought as deeply about the alternatives to a nuclear-armed world as Jonathan Schell, who concluded that "the days when humanity can hope to save itself from force with force are over . . . force can lead only to more force, not to peace." We have no realistic choice but to abolish nuclear weapons and the idea that a decent and stable world order can be founded on "the threat to extinguish humanity." Abolition of nuclear weapons, he writes, would be "an indispensable though insufficient recognition by the human species of the terrible, mortal predicament it has got itself into and a concrete expression of its resolve to find a solution." Schell also regarded the abolition of nuclear weapons as a foundation for dealing with climate change. It would create the habits of cooperation among nations, release large amounts of money otherwise squandered on unusable weapons, and free the Earth from a mortal danger. It would also liberate the creative energies and spirits of some of the best scientists who would no longer be employed to make heinous weaponry.[65]

Predictably, critics of disarmament will say that the world of the twenty-first century is a dangerous place and cite the long list of some indisputably "bad guys" from ISIS to Vladimir Putin, to yet another enemy du jour. And sometimes they are right. The problem, of course, is that others often see us as a threat, too. And sometimes they too are right. On both sides problems arise due to how humans perceive reality while under emotional stress and how they come with the illusions of doctrine and pathologies of nationalism. Under stress, clarity of mind and good judgment are casualties of "group think," "self-fulfilling prophecies," "denial," "terror management," and plain error. And as economist and systems thinker Kenneth Boulding once remarked, one should "never underestimate the power of stupidity." Neither should we forget its first cousin, ignorance—including the willful ignorance of how we are perceived by others.[66]

The abolition of nuclear weapons, moreover, must occur in tandem with a larger effort to eliminate violence from our security policies. Violence in our time is self-defeating and self-perpetuating. It exhausts treasuries, consumes our creativity, slaughters the innocent, rewards humanity's worst traits, and leaves a trail of human, material, and ecological ruin in its wake. "War," in Christopher Hedges's words, "is a drug . . . it exposes the capacity for evil that lurks not far below the surface within all of us." In a world of calamitous Earth-destroying weaponry that is widely dispersed and casually, perhaps intermittently, controlled, there will be no such thing as victory in any war. Institutionalized violence is a form of mass insanity and if allowed to go on much longer, sooner or later it will destroy us.[67]

The transition to a nonviolent world will not be easy. The habits of violence and war are deeply etched into our behavior, politics, economy, history, and perhaps our genes. The military hero is greatly celebrated, but heroism in all of its other less testosterone-driven forms gets seldom so much as a whisper of appreciation. I refer to the daily, mostly unnoticed heroism of social workers, teachers, coaches, public officials, working men and women, fighters for justice, tree planters, gardeners, and peacemakers in tough places. It is true that the world of sovereign nation-states and groups of terrorists, jihadists, revanchists, and the deluded *is* a dangerous place. If we intend to survive through the long emergency ahead, however, we must learn the disciplined arts of peaceful conflict resolution and nonviolence as well as intelligent ways to avoid conflict-generating situations, including our dependency on resources beyond our borders. Advocating for peace is the only practical option left to us. And as William Penn once said, "Somebody must begin it."[68]

As difficult as the transition to a nonviolent world will unquestionably be, we already have many of the tools for "getting to yes" without shedding blood. We have rules, regulations, laws, and even stoplights at traffic intersections to mitigate damage and conflict. We have "time out" periods in schools, cooling-off sessions in counseling, mediation in labor disputes, courts of law instead of duels, and penalty boxes in hockey. There is, further, a large and growing body of experience with nonviolent conflict-resolution strategies; consider those applied successfully by Mahatma Gandhi, Martin Luther King, and others, all of which make manifest the power of the powerless. What remains is the long work of education, training, and institution-building necessary to change the larger culture. Conflict resolution should be a prominent part of education from kindergarten through graduate school. It should be an important major for students in higher education, not an afterthought. Courses in conflict resolution should also be required of all public officials. It should be a prominent part of our public dialogue and political debates. Those trained as specialists in violence should also be well trained in the science of peace-making. And peace academies ought to flourish everywhere at public expense because they are in the public interest. Conflicts of interest and perception will always be difficult to resolve, but they need not lead to violence. The difference will test our determination to consistently apply the methods of conflict resolution in order to build the capacity of international courts and agencies of international peacekeeping. As with any significant change, we will learn the art of sustainable peace by taking small steps with our eyes on the prize: a world rid of nuclear weapons and with a growing institutional and emotional capacity to reconcile conflicts and competing claims without resorting to violence.[69]

A sustainable economy cannot long exist as an island in a society and global system ruled by threats, violence, and war and the profit made from such things. On some apocalyptic day, something will go horribly, horribly wrong. In the meantime, the war system sucks everything dry, bankrupting better possibilities and desiccating the practice of democracy. The language of war and conquest corrupts our habits of speech and frames our worldviews. It gives license to our worst traits. It thrives in the darkness of fear and ignorance. The habits of violence drain imagination, veto better possibilities, and squander people and treasure alike. Our military weapons glisten ominously in their terrible power while our schools decay and our infrastructure rusts. Not least, the mindset of violence shapes our perceptions of nature as merely another thing to be conquered. But there can be no sustainable and just economy founded on the idea of harmony between humans and natural systems in a society preoccupied with violence, threats, and war.[70]

Assuming that we can wean ourselves from our addiction to violence and war, what can be said about the next economy? Of economic predictions there is no end. They tend to range between euphoria of a commercial culture and the kind of despair befitting a dismal science. At the euphoric end, Jeremy Rifkin foresees a world in which

> The Internet of Things will connect every thing with every one in an integrated global network. People, machines, natural resources, production lines, logistic networks, consumption habits, recycling flows, and virtually every other aspect of economic and social life will

be linked via sensors and software to the IoT platform, continually feeding Big Data to every node—businesses homes, vehicles—moment to moment . . . [culminating in] a coherent operating network, allowing every human being and every thing to communicate with one another in searching out synergies and facilitating interconnections in ways that optimize the thermodynamic efficiencies of society while ensuring the well-being of the Earth as a whole.[71]

Privacy? Not to worry, it's not an inherent right and anyway it's overrated. Hackers? Well, that is a worry. The IoT is a banana forest for hacker monkeys, but we'll figure that out later. Vulnerability of the electric grid? Yeah, that's a problem. Total government surveillance 24/7 that would make Orwell's *1984* look like a New England town meeting? Silence. Capitalism? Well, it goes away, mostly. We will make our own stuff with three-dimensional printers powered by solar collectors at zero marginal cost. Oh by the way, late in the book we learn that climate change could wreck the IoT party and that there are "cyberterrorists out there." But the connections between the mindset that would wreck the planet and that behind the IoT remain unexamined and unexplained.[72]

Rifkin's view, as well as that of the leaders of Google and Facebook, begs the questions of how we connect what to what and for what purpose. Are there firebreaks in the system? Are there things that should not be connected? Are there data that should not be collected? Who controls the Big Data that are so easy to collect, store, and use to whatever purpose? Who controls the use of data and who controls the controllers? Is our goal to make the world safe, cheap, and solar powered for "proconsumers" who are AWOL as citizens and incapable of thinking critically about our prosthetic technologies? Do we intend to surrender our sense of direction to a GIS or our capacity to open a window, detect the weather, or remember anything on our own to smart sensors who will do it all for us? Does the IoT along with dependence on Google make us dumber? Beyond some unknown point, the smarter, networked, wired up, electronic version of the "Grand Inquisitor" becomes a tyrant, not a benefactor. What happens to income inequality, power, and justice? For all the hype, the IoT is still a world of hyperconsumption and human domination but it has been outsourced to smart machines over which people will have less and less control. And where will they find the silence and solitude to restore hearts and souls then?[73]

Other grand visions of the future economy are not much more satisfactory or convincing. The reason is that the future is less predictable than ever. The rate of change is too rapid, the complexity too vast, the possibilities too many and too disjunctive. But we do know that present economic arrangements are not suited for the long emergency and will have to be remodeled or scrapped altogether. This is a systems crisis and can only be solved by changing the economic and political systems and rebuilding their intellectual foundations.[74]

And that's the heart of the problem. We presently lack both the vision and the means adequate to change large systems to good effect toward reasonably predictable ends. The problem is that systemic change happens either slowly by accretion or suddenly and chaotically when things collapse. And when collapse occurs, the results are often unpredictable and sometimes tragic. Systemic, top-down change in the economy and society is today, as it has always been, very difficult and is one reason for the stranglehold that the wealthy and powerful have on governments and the media. So what's to be done?

Society is a partnership not only between those who are
living, but between those who are living, those who are dead,
and those who are to be born.

—Edmund Burke

We can't go on living like this.

—Tony Judt

In concluding his magnus opus *The Myth of the Machine*, Lewis
Mumford described what he regarded as the only possible pathway
to progressive change:

> Though no immediate and complete escape from the ongoing power
> system is possible, least of all through mass violence, the changes that
> will restore autonomy and initiative to the human person all lie within
> the province of each individual soul, once it is roused. Nothing could
> be more damaging to the myth of the machine, and to the dehuman-
> ized social order it has brought into existence, than a steady with-
> drawal of interest, a slowing down of tempo, a stoppage of senseless
> routines and mindless acts . . . the long buried seeds of a richer human
> culture are now ready to strike root and grow . . . the gates of the tech-
> nocratic prison will open automatically, despite their rusty ancient
> hinges, as soon as we choose to walk out.[1]

That is a plausible description of our situation in the first half of the
twenty-first century: gridlock, vacillation, and corruption at the top
but vibrant creativity and innovation increasingly evident at local, city,
and regional levels. Yet sooner rather than later major policy changes
are needed at the state and federal levels to speed the transition to a
renewable energy economy, build twenty-first-century transportation

and communication systems, restore fairness in the distribution of wealth, protect and expand voting rights, regulate banks and financial institutions, separate money from electoral politics, repair flaws in the Constitution . . . the list of items on our deferred maintenance list is long. Opposition preventing reforms has been strongly entrenched in Congress, corporations, the Supreme Court, and the media.

That is not the whole story, however. We may also have reached the limits of continental-scale governance as James Madison once feared. Madison, the principal architect of the U.S. Constitution and its most important defender, was unsure that it could hold together against the powerful centrifugal forces of faction, population growth, and geography. But the Union did hold together through the travails of Civil War, two World Wars, Depression, and the Cold War, much longer than Madison thought possible. Today, however, we have reached an impasse. Reforms have been thwarted by the combined effects of geographic scale, regional differences, the complexity of the problems, an entrenched oligarchy, ideological differences, growing inequality, and flaws inherent in the Constitution—all of which will be amplified by population growth and climate change. The tide of unsolved problems facing society is rising, but the federal capacity to act is being deliberately weakened or thwarted. Intermittent reforms have been partial, starved for funding, and then tied up in endless lawsuits. It is time to rethink and reform the basic political economy of democracy in the United States.[2]

To that end, Mumford proposed no massive uprising or movement, only "a steady withdrawal of interest," which presumes that we would find what previously had held us in thrall and made us dependent was neither interesting nor important. Imagine, for example, if we no longer thought Coca-Cola and its competitors selling similarly over-caffeinated sugar-saturated water in throwaway plastic bottles interesting, important, or even thirst quenching but we did find our local water systems interesting, important, and a superior way to quench thirst without causing tooth decay and obesity—and therefore worth protecting from contamination and corporate ownership.

Imagine discovering that cereals with names like "Count Chocula" were neither interesting nor edible, but that a local farmer selling eggs from free-range chickens and granola from organically grown healthy local grains could make breakfast worth eating. Imagine discovering that ExxonMobil's gasoline was no longer interesting or important because it was no longer necessary to power your electric car that was recharged by a solar system on your roof. Further, imagine discovering that you no longer needed even an electric car because local bike trails connected you better to your community with the added advantages of improved heart, lung, and mental capacity. Imagine discovering that fracking was unnecessary, uninteresting, and unprofitable because improving the energy efficiency of our homes and businesses is a cheaper, easier, and faster way to reach affordable energy security while preserving community rights and protecting public health. Imagine discovering that your electric utility was no longer important or interesting because the combination of improved energy efficiency and on-site solar electric systems made their coal-fired, climate changing, mountaintop-removing, mercury-dispensing electricity neither interesting nor necessary. Imagine discovering a nearby farmers' market selling fresh, locally grown fruits, vegetables, grass-fed beef, and free-range chickens that rendered the chemically saturated, engineered products of Monsanto, Cargill, General Foods, Smithfield, and their few competitors considerably less interesting and important. Imagine discovering that conversation at the local pub over locally crafted beer was much more interesting, honest, and convivial than listening to whatever it is that is dispensed by Fox News, the angry eruptions of Bill O'Reilly or the convoluted logic of Sarah Palin, or the vitriol of Ann Coulter. Imagine a local community with ecologically competent gardeners, farmers, builders, repair technicians, artists, writers, makers, entrepreneurs, and teachers more interesting and prosperous than the shopping mall economy and the will-o'-the-wisp fantasy of Big Box prosperity funded by footloose capital from absentee investors. Imagine people waking up to the idea that locally owned enterprises that keep money in the local economy longer were more interesting, profitable, and resilient than stores owned by corporate behemoths that siphon money and good work out of the community.

Imagine, as Mumford proposed, slowing the frenetic tempo of the treadmill that keeps us moving, distracted, searching, consuming, and agitated, which in turn makes us too busy to think. I do not believe that is an accident. Rather, I think educator John Taylor Gatto is right in saying that "mass dumbness is vital to modern society." Dumber people are better suited for servitude because they have been told they are free and lack the wherewithal to see their fetters. Slowing the tempo gives us time, if we choose, to observe, reflect, think, meditate, and perhaps stop the "senseless routines and mindless acts" that deaden our souls, steal our days, and diminish our common wealth.[3]

Mumford's withdrawal and escape from the "technocratic prison" requires no Congressional or federal permission, only the willpower and community-scale organization of competent, informed, and motivated citizens. He proposes not so much the devolution of government powers as the development of practical capabilities and infrastructure at the local, community, and regional levels so that the need for many centralized services and administration is lessened. The result is a Declaration of Independence from Big Oil, Big Pharma, Agribusiness, Big Utilities, Big Finance, and Big Media, by people who have become citizens and neighbors able and willing to take back those parts of their lives closest to them such as growing food, energy production, shelter and community design, transportation, finance, and economic livelihood. Every garden, solar collector, bicycle, energy efficiency improvement, locally owned business, co-op, self-designed home, "no buy" day, recycling center, community baseball league, walk in the woods, time spent with kids, and neighborhood bake sale is a Declaration of Independence from distant powers as well as a Declaration of Interdependence with our neighbors and our places. In time, a steady withdrawal might grow, as Mumford thought, into that "richer human culture," and the doors of the technocratic prison would open and perhaps cascade into a society-wide political and social transformation.

Updated to the realities of the long emergency, Mumford's vision is not an argument against government, but rather an acknowledgment of its limits in an age of political dysfunction and plutocracy. I think it neither conservative nor liberal in outlook. It is not anti-capitalist, but it would be foolish to overlook the pathologies of unregulated, global-scale capitalism on one hand or the possibilities for an innovative, human-scaled, accountable, and transparent capitalism on the other. It is not an argument for socialism, but it would be foolish as well to overlook the need for greater social solidarity in the centuries of the long emergency. It is, in fact, consistent with the logic of federalism and the principle of subsidiarity that aims to assign functions as much as possible to the lowest level of government and closest to those immediately affected. In so doing, it aims to lessen or remove altogether those functions and problems that overburden government agencies, or those that cannot be resolved at the national level, but might be solved or lessened at lower levels. It is both Hamiltonian in the acknowledgment that we need a competent and accountable central government that conducts the public business with energy and foresight, and Jeffersonian in the hope for a thoughtful, informed, and competent citizenry of property owners with a personal stake in the society.

A competent, energetic, and accountable national government is necessary to uphold the scaffolding of laws and enforcement that maintains public safety, protects common wealth, and preserves the economic and moral ligaments that bind us together. In times of emergency we also count on national leaders to summon the angels of our better nature, remind us of our highest values, raise our sights to a farther horizon, and call us to do our duty. America was born in opposition to what was deemed to be tyranny of King George III and the English Parliament, but the founders devised a system that made change, however obvious and necessary, difficult and slow and governance awkward and clumsy. The fear of democracy permeated their august deliberations, so they erected barriers against possible excesses committed by "we the People." In the years and centuries ahead we will surely need an effective, accountable, energetic, and democratic national government, but with a difference. It must provide the larger framework of laws, regulations, and security that allow for regional and bioregional differences and that energize policy innovation and public engagement at the grassroots. In the former role the concerns are those of justice, equity, security (broadly defined), prosperity, goal setting, law enforcement, regulation for the common good, and foresight. In the latter, the intention is to enhance citizen engagement, administrative flexibility, and the innovative execution of policy goals. The layering and division of authority, however, cannot be used as a pretext to evade the law or dilute standards in the service of baser purposes.

The devolution of some functions and capacities to lower levels, however, is not a magic solution. As social philosopher Leopold Kohr once observed: "Neither the problems of war nor those relating to the purely internal criminality of societies disappear," but at a smaller scale "they become soluble . . . *Every* vice shrinks in significance with the shrinking size of the social unit in which it develops . . . Indeed, there is no misery on earth that cannot be successfully handled on a small scale as, conversely, there is no misery on earth that can be handled at all *except* on a small scale. In vastness everything crumbles, even the good, because as will increasingly become evident, the world's one and only problem is not wickedness but bigness."[4]

Succession movements in Canada, Spain, Scotland, and elsewhere testify to the centrifugal pressures that seem only to grow for reasons other than mere contentiousness. There is a powerful logic at work in decentralism and we would do well to work with it.

Further, a renaissance of local democracy is a goal on which both liberals and conservatives can agree. Conservative economist Frederick Hayek, for example, once wrote:

> Nowhere has democracy ever worked well without a great measure of local self-government, providing a school of political training for the people at large . . . It is only where responsibility can be learned and practiced in affairs with which most people are familiar, where it is the awareness of one's neighbor rather than some theoretical knowledge of the needs of other people which guides action, that the ordi-

nary man can take a real part in public affairs because they concern the world he knows.[5]

Conservative political scientist Frank Bryan similarly calls "real face-to-face democracy . . . the oxygen of our representative Republic." From a liberal perspective in a different time, John Dewey wrote that "democracy must begin at home, and its home is the neighborly community." Historian and economist Gar Alperovitz adds that "there is no way to rebuild Democracy . . . without nurturing the conditions of democracy with a small d in everyday life—including the economic institutions." It is in the human-scale settings of school, neighborhood, and community where people first and best learn the habits of democracy and the rights and duties of citizens. At this scale they learn the art and power of reasoned civic and civil dialogue, engagement with issues, the necessity of compromise, and the limits of ideology. In their important study *Size and Democracy*, political scientists Robert Dahl and Edward Tufte note that "very small units seem to us necessary to provide a place where ordinary people can acquire the sense and the reality of moral responsibility and political effectiveness." But what is the necessary experience of democracy?[6]

Susan Clark and Woden Teachout conclude that real democracy "requires that people understand each other . . . only then can they talk clearly and figure out the best decision—one that will last." The process of democratic deliberation, in other words, has its own pace that cannot exceed the speed limits of human dialogue, the time required to reason together, and to reach compromise. It cannot be made efficient as that word is commonly understood. It takes time to understand other perspectives, sort out complexities, find workable compromises, and reach durable solutions.[7]

In this regard, the Vermont town meeting is an important laboratory for both the practice and study of democracy in small-scale settings. It is said to convey civic skills such as running a meeting by *Robert's Rules of Order*, to be "a tutor for mediating values," and to teach forbearance in the face of one's own and others' intolerance. It has worked well, by most accounts, because the scale is small, the purposes broad, towns are free to make mistakes on significant issues, and the process helps people to master the mechanics of pragmatic, problem-solving governance that leaves little room for ideology. Democracy, in other words, becomes the default. There are two other factors worth noting. Vermonters, by and large, are joined to the land as farmers, foresters, and otherwise ecologically attentive and practical people. Moreover, the economy is predominantly one of owner operators, not multinational global corporations. The result is a high degree of efficacy: people believe that their opinions, efforts, and votes really matter.[8]

But democracy in Vermont and elsewhere mostly stops at the factory door or just outside the C-suite offices in corporations that are run as undemocratically as any authoritarian state. The result is schizophrenic. We describe ourselves as a democratic society, but spend our days working in organizations that are ruled otherwise. Moreover, we don't seem to mind the gap. Is there a case for extending democracy into workplaces and corporate management? Yale political scientist Robert Dahl, perhaps the greatest student of democracy in the past century, believed that democratizing economic enterprises would neither violate the rights of private property nor necessarily impair its performance. He asked,

> If democracy is justified in governing the state, then it is also justified in governing economic enterprises. What is more, if it cannot be justified in governing economic enterprises, we do not quite see how it can be justified in governing the state.[9]

If broadly implemented, the extension of democracy within the corporation, a step that seems so radical, would in time become merely normal and afford many of the same advantages that genuine participation brings to the conduct of public affairs. But would thoroughly democratic enterprises make better decisions than those made by overpaid and often isolated executives perched at the top of corporate hierarchies? Would those decisions be more honest or transparent? Would the elected executives of democratically run companies be as likely to cheat on quality, financial statements, or pollution regulations, or incur the risks of destabilizing the climate? I cannot say that larger, democratically selected groups of people are necessarily more competent, honest, or virtuous than members of very small management groups. But they would be less isolated by background, hierarchy, and the presumptions of privilege. They would exist in a wider circle of responsibility where it would be harder to fudge numbers or make bone-headed or stone-hearted decisions without someone saying "I disagree." Diverse backgrounds, training, and lives would more likely be reflected in the collective manner of thinking and might render morality more expansive and less malleable to the demands of the quarterly report. They will not immunize groups from "groupthink," or the many other pathologies of collective decision-making, but they do make them less likely and, I think, more reparable when they do occur.

Finally, democracy as it exists in both theory and practice does not extend to the actual decision-making process. "Even participatory democratic theory," in Beth Simone Noveck's words, "fails to identify concrete institutional opportunities for genuinely substantive participation in *formal* decisionmaking processes." She proposes a "new legal framework" that opens new pathways for public participation in policy-making. "The conflict between expertise and democracy," she writes, "is artificial," a vestige of the industrial revolution, Newtonian science, and Progressive-era belief in expertise. She proposes to harness the internet and social media to identify expertise increasingly diffused throughout society and "bring participatory practices to governance, including administrative agencies, by expanding the role for citizen experts and diversifying the conception of engagement." The effects of broader participation on actual administrative decision-making could be important in several ways. Chiefly, such participation could help break down the proverbial silos of professional expertise and broaden the awareness of possibilities that cross conventional lines and thereby improve policy outcomes. In short, Noveck argues plausibly that connecting intelligent people with actual decision-making would make the state smarter and deepen the practice of democracy.[10]

The biggest obstacle to seeing such possibilities, however, is the poverty of our public and political conversations, which have become narrow, unimaginative, and constricted. The public arena has come to resemble a yard with an electronic fence that keeps the pet dog obediently confined. Trespasses are painful and can be fatal to the boundary-crossing ambitions of roving dogs and adventuresome people alike. In the conventional wisdom, we can talk and presumably think only about the corporate-dominated capitalist economy as it presently exists, not as it could be. We are permitted to think of wealth only as exclusively owned and exclusionary, not as common wealth. We are permitted to think of politics as it presently exists in a radically unequal society dominated by a handful of oligarchs, not as a collective liberation from fear, oppression, and injustice. We are conditioned to think of political dialogue as entertainment that appeals to our baser selves, not as a way to advance the common good and raise our collective IQ. We are permitted to think of justice as beneficence bestowed, not as an inalienable right. We are permitted to think of humans as wholly and rightfully dominant over the Earth and its animals, lands, waters, and mysteries, but not as participants in the fellowship of life.[11]

There are more promising options available, however. As Gar Alperovitz, Gus Speth, Michael Shuman, and Jeff Gates have shown, we can reform existing institutions and develop better ones that are rooted in local democracy. And if capitalism is a good thing, then we need more capitalists, not fewer. We need a new deal between capitalism and democracy that distributes wealth more broadly, reduces inequalities, democratizes the workplace, and protects nature. We need to extend public ownership of common property—those assets that should not be privately owned such as water systems. We will need new institutions, new ways of thinking, new paradigms, better ways of defining the relationship between the public and private realms, and new ways of defining the rights of property and ownership. In Alperovitz's words:

Without local democracy, there can be no culture of democratic practice; without security and time, there can be only a weak citizenry;

without decentralization, it is difficult to mobilize democratic practice and accountability; and without major and far-reaching new forms of wealth-holding, there can never be adequate support for the conditions and policies needed to build a more egalitarian and free democratic culture.[12]

We need an economy that serves all of the people and is held accountable to a truly open democratic political system. The founders of the United States regarded their work as a design project. The long emergency presents us with an even larger and more urgent design challenge: to join radical democracy with better, cheaper, and ecologically smarter ways to provision ourselves with food, energy, transportation, and information that liberate us from the persistent sources of oligarchy while building local competence.[13]

Design is all the rage. We wear designer jeans, live in designer houses, drive cars with designer interiors. We pay "starchitects" sizeable fortunes to design buildings that defy gravity, conventional geometry, urbanity, and human scale. Some are preparing to help us redesign our genes by means liberated from sexual selection and unhindered by philosophy or the piddling constraints of conventional morality. The goal is superhuman perfection (though the result could be the geneticists' version of an arms race). Others are designing the means to make consumption easier and faster by tailoring advertisements to our electronically recorded habits and desires. The goal is to design consumption patterns form-fitted to our particular person in order to make higher profits and greater dependency. Whatever you want, someone will provide it or design it for you, but at a price high enough to exclude the great majority of humanity and without any thought of due process for those who will follow us—future generations who will find their grasp on life, liberty, and happiness more difficult.

Missing in the froth of new possibilities in a brave new electronic world are flesh and blood people and their connection to the design choices being made, ostensibly, for them. The poor and disadvantaged are conspicuously absent from the articles and advertisements in air-brushed architectural and design magazines. Missing, too, is a thoughtful analysis of the fine print of the deal, and an acknowledgment of human fallibility and ignorance. What are the social, ecological, and spiritual implications of various design choices? How might these affect other species and future generations?

The ironies stack up like cordwood. For one, we pay lots of money to visit the historic cities of Europe precisely because their cramped old towns with crooked, narrow streets and ancient buildings retain much of their charm and attraction as human-scaled places to live and work. They grew incrementally, as Jane Jacobs once explained, to meet particular needs in specific places, not as the result of the grand linear schemes of designers like Robert Moses and Le Corbusier with their straight lines, motorized transportation systems, and efficiencies tailored to the needs of the hurried class.

No reasonable person can deny the need for intelligent design, particularly to the extent that it reduces material, water, and energy use, and therefore pollution. But since World War II we have discarded much of the older design intelligence and infrastructure. I recall, for example, the light-rail system of Pittsburgh that was once an efficient, dependable, and cheap way to move around the City. Oberlin, Ohio, my present home, once was a part of an interurban system that connected all of the cities and towns between Toledo and Cleveland. Light-rail systems like these were dismantled all over the United States not because they were inefficient or unpopular, but because companies like General Motors bought them up in order to put them out of business. We now drive their cars, with umbilical cords that stretch to depleting oil fields in distant places where we are not much appreciated. We also spend much of our lives caught in traffic jams, burn up our emotional energy in road rages, drive through award-winning ugliness, breathe air contaminated with whatever comes out of our tailpipes, and risk being among the hundreds of thousands killed or injured on the highways, all while paying exorbitant taxes to protect our access to other peoples' oil. No reasons were given; no vote was ever taken. The market, it was said, had spoken, but in truth the public good had been hijacked by corporate pirates in three-piece suits. Advertisers, who had long before been mobilized to sell cars and their accessories, joined in the hijacking and geared up for the post–World War II boom that drove our economy for the next half century by selling chrome-trimmed fantasies, images of higher social status, roadside wonders, fast foods, and the promise of excitement just over the next horizon.[14]

In this case the political economy of design was neither political in the sense of being the result of a public decision reached through democratic deliberation, nor economic in any honest way. Any true accounting of the full costs of our automobile-driven culture also would have to include the role of ExxonMobil and other oil companies in delaying action on climate change, which they knew to be a reality as far back as 1977. How many people will die because they lavishly funded four decades of climate denial? How many trillions of dollars of damage might have been avoided had they chosen to help in the transition away from the fossil-fuel-powered system they created? A great deal of the debt for that half-century binge of individualized, oil-dependent, land-eating, polluting, and democracy-corrupting system will be passed on to our descendants. There will have to be an accounting of sorts, but probably neither truth nor reconciliation because we do not have the laws or institutions to adjudicate lethal malfeasance at this scale.

All design, in other words, exists in a larger framework of political economy by which costs and benefits are distributed within society and across generations. Whether we acknowledge it or not, the failure to design with physical and ecological realities incurs irrevocable costs as well. Commercial and industrial designers, however, mostly prefer to regard their work either as politically and ethically neutral or merely as a matter of aesthetics and novelty. Steve Jobs, for one, learned how to make computers that were considerably more than tools for computation and communication: they were designed to light up the pleasure centers of the brain. Jobs, in Sue Halpern's words, used "enchanting theatrics, exquisite marketing, and seductive packaging to convince millions . . . that the provenance of Apple devices was magical, too." Lost in the hype and cultish aspects were the lives of underpaid, exploited workers and the mountains of electronic trash thrown out to buy next year's I-whatever, which Halpern muses "may be the longest-lasting legacy of Steve Jobs' art." As Jobs understood, design has powerful effects on the consumer.[15]

It is a fact well understood throughout the business. "Once people come in," as a software engineer at Instagram says, "the network effect kicks in . . . then it takes on a life of its own." The goal is to make it "enthralling" and "difficult to put down"—in a word, addictive. Every person so addicted faces "1,000 people on the other side of the screen whose job is to break down the self-regulation that you have."[16]

Frank Lloyd Wright once bragged that he could design a house for a newly married couple madly in love that would cause them to divorce in a matter of weeks. But Wright, the architect, would have been able to manipulate only materials, geometry, space, and landscapes. Contemporary designers work with much more powerful tools and sometimes with bad effects.

In *Addiction by Design*, Natasha Schüll describes the use of design to turn gamblers into gambling addicts. One gaming-machine designer, for example, "really knows how to get into the head of a fifty-year-old woman and figure out what she wants." The casino owners who buy his machines only want her money, but he knows that her purse is connected to specific parts of her brain that can be manipulated. "Gambling machines" like those he designs "are complex calculative devices that operate to redistribute gamblers' stakes in a very precise, calibrated, and 'scientific' way . . . [they] operate as vehicles of enchantment, galvanizing what Zygmunt Bauman has described as 'human spontaneity, drives, impulses, and inclinations resistant to prediction and rational justification,' or in Weber's words, 'irrational and emotional elements that escape calculation.'" Such machines work in the "fashion of psychostimulants, like cocaine or amphetamines. They energize and de-energize the brain in more rapid cycles." In the words of one gambler, "You're in a trance, you're on autopilot . . . the zone is like a magnet, it just pulls you in and holds you there." As long as you have money. "It's our duty," as the CEO of Las Vegas Stratosphere puts it, "to extract as much money as we can

SUSTAINABLE DEMOCRACY?

from customers." The entire design of the modern casino aims to create a womb to keep gamblers oblivious of everything else, until their wallets have been drained. "It is not uncommon," Schüll writes, "to hear of machine gamblers so absorbed in play that they were oblivious to rising flood waters at their feet or smoke and fire alarms that blared at deafening levels . . . or even a dying man at their feet." So addicted, gamblers will stay at it for twelve hours or longer, ignoring all bodily needs to satisfy the craving.[17]

The gambling industry described by Schüll resembles in many ways the larger world of modern advertising that drives consumption. At least since Edward Bernays created the prototype for the modern advertising firm, capitalists have intended to make us dependable and dependent consumers, that is to say consumption addicts by design. The sellers often know more about us than we know of ourselves. They track our behavior—bodily responses to various stimuli, eye motions, and our every flinch and fantasy—trolling for any information that might be useful in order to sell us more of what we don't need but can be manipulated to crave. They aim to orchestrate our fears, phobias, and desires and shape us into more ardent consumers.

The art and sciences of design, in other words, constitute a set of powerful tools drawn from the disciplines of psychology, neuroscience, sociology, anthropology, computer science, architecture, and interior design. These can be used to manipulate, deceive, and render dependent. Alternatively, they could be used to help us undo our addictions and mindless consumption and build a convivial, democratic, fair, decent, healthy, pollution-free, and nonviolent world powered by renewable energy and populated with people made competent by design. There is an important distinction between conventional design, which is simply the making of things to appeal to status needs and transient fashion, and ecological design, which is the skill to make things of value that last and fit harmoniously in their ecological, cultural, and historical context.

Ecological design aims to calibrate human actions with the way natural systems work as particular places, larger landscapes, and whole

197

ecologies. It aims to work with, not against, the flows of energy and natural cycling of materials. It aims to conserve and regenerate the basis for life and human flourishing. Architect Stuart Walker notes, "If design is to more effectively address sustainability it has to transcend utility and conventional function-led, and especially technology-led approaches." He calls on designers to rise above "the calculated creation of dissatisfaction" and to "think more comprehensively about the products we already produce and their implications." Design, in other words, must be an act of integration, not just specialization, with the goal of creating wholeness that includes our spiritual well-being. In Robert Grudin's words, design, "unlike any other concept . . . calls for us to create a unity of part with whole, a concord of form and function, a finished product that is harmonious with society and with nature."[18]

By that standard, there are some things that are discordant and should not be done. And the distinction between what designers can do and what they should do requires a statement of design ethics to inform professional conduct rather like the Hippocratic Oath in medicine. Engineer M. W. Thring, for example, proposes a standard that includes all of the consequences of engineering and design, in-cluding those affecting "the subjective qualities of human life such as self-fulfillment, happiness, inner freedom and love." Specifically, designers should see all of their work, as engineer Seaton Baxter puts it, as a manifestation of "co-evolution with the natural systems of the world," increasing the quality of life while reducing the consump-tion of resources.[19]

The basic rules for ecological design are to

- maximize the use of solar energy;
- protect diversity of all kinds;
- eliminate waste;
- use nature as the model;
- make it affordable;
- design for repair and disassembly;
- build in redundancy and resilience;

- maximize public participation; and
- make it beautiful.[20]

Design, in other words, is a healing art in the broadest sense of the word. It is no accident that the word health is related to the words holy, whole, holism, and healing. Neither, I think, is it an accident that the root words for religion and ecology imply wholeness and connection. Something that is ecologically designed should embrace the health of people, ecologies, and social institutions as interacting parts of a larger whole.

Further, good design should increase the reservoir of practical ecological competence at the local level, allowing us to do more for ourselves and for each other—much as we once did as competent people, good neighbors, and active citizens. Excessive dependency has made us less intelligent in many respects. Our smartness has migrated upward into our technologies and systems of technology, which aggregate our considerable collective intelligence but leave us as individuals clueless and stupid. Few, if any, know how these remarkable technologies are made or how they work, let alone how to live without them. Leonard Read's 1958 essay "I, Pencil" illustrates the point by showing that not a single person on Earth knew how to make such a simple thing as a lead pencil, or almost anything else. It is possible, in other words, that our individual competence, and perhaps even intelligence, is inversely proportional to the brainpower that goes into our technologies and systems of all kinds—all designed by teams who each know a small part of the final ensemble but little of the larger system of which it is a part or the effects it may have on the wider culture.[21]

Ecological design, second, should help ground us. It should inform us of where we are and the ecology and energy flows that sustain us in that place. In a world where any one location looks much like any other, we have lost sight of the costs and consequences of the systems that provision us so seamlessly with food, energy, and materials, and are oblivious to their inherent fragility. Good design should help us know where we are and how to be ecologically competent in that place. Those places should be ecologically designed to retain water for drought periods, manage flood waters, grow food and fiber, sustain wildlife, and absorb carbon. They should be working landscapes that blend agro-forestry, mixed-use permacultures, intensive agricultural and gardening zones, viticulture, aquaculture, water purification, and recreation. And they should be managed by local citizens and used to train young people for lives of growing the essentials in managed, integrated ecologies.

Third, ecological design should enhance the opportunities for conviviality, celebration, and face-to-face democracy. Communities with front porches, public squares, community gardens and solar systems, neighborhood stores, corner pubs, and open places of worship are more likely to thrive in the long emergency because they foster neighborliness, community cohesion, and support systems in good times and bad. Good design will engage people in the making of their homes, neighborhoods, towns, and regions, thereby increasing our civic intelligence and our joy in life. Designers in this way are facilitators in a larger public conversation and architects of better possibilities, not just makers of buildings and things.[22]

Fourth, good ecological design should improve the resilience of a community in the face of larger storms, longer droughts, and higher winds. Properly designed urban infrastructure would mimic the ecological functions of forests, grasslands, and shores and eliminate much of the costs of concrete hardscaping, pipes, and water treatment systems. Infrastructure designed to serve multiple functions across the conventional jurisdictions and sectors can, in Hillary Brown's words, "produce ongoing cost economies, reduce disruption, and conserve valuable resources."[23] What she calls "next generation infrastructure" is a call for smarter, less expensive systems that let natural systems do much of the work of water storage, purification, and carbon sequestration in ways that enhance property values while improving opportunities for recreation and urban agriculture.

The long emergency will add to the design challenges ahead. Designers must reckon with a world of higher temperatures, stronger winds, more frequent and larger storms, rising ocean levels, longer droughts, much larger rainfall events, and new diseases. These could cause interruptions in supplies of food, energy, and water, which in turn may trigger social disruptions. We must design our communities with full awareness of the fragility of the present world, as Jared Diamond and others warn, and anticipate how to both endure crises and recover from them.[24]

Rapid climate change puts a great deal of conventional design into question. Taller buildings are vulnerable to wind shear, electricity outages, and terrorists. Sealed buildings work only as long as HVAC systems and complicated electronic controls work, which means they also require uninterrupted electricity, expert managers at hand, and dependable supply chains. Instead of more engineering virtuosity, then, designers should "start re-engineering our aspirations, infra-structures and lifestyles for a softer landing in a post-fossil [fuel] world."[25]

Beginning in neighborhoods, towns, and cities, the great work of our generation is to make that softer landing in a post-fossil-fuel and post-consumer economy possible—and in ways that are prosperous, fair, durable, resilient, convivial, and democratic. The new economy will have to be powered by renewable energy and both recycle and reuse its wastes. Of necessity it will be much more focused on essentials of food, energy, shelter, clean water, education, and the arts, and rooted in its place and bioregion. It will be built by local people who cherish and understand their places and the place of nature in a sustainable economy. But it must be a political economy, a product of revitalized grassroots as well. If it is to flourish, it must grow out of the union of ecological competence and the practice of radical democracy.

eleven

CITIES IN A HOTTER TIME

A system is anything that is not chaos. . . . any structure that
exhibits order and pattern.

—Kenneth Boulding

A system is an interconnected set of elements that is
coherently organized in a way that achieves something. . . . [it]
must consist of three kinds of things: elements, interconnections,
and a function or purpose.

—Donella Meadows

A system [is] a set of units or elements interconnected so that
changes in some elements or their relations produce changes in
other parts of the system, and . . . the entire system exhibits
properties and behaviors that are different from those of the parts.

—Robert Jervis

One of the most important ideas in modern science is the idea of a
system; and it is almost impossible to define.

—Garrett Hardin

In a deep southern accent the caller said: "David Orr, this is David Crockett from Chattanooga." He spoke as if we were long-time
friends or at least family members. Thinking this was a gag call, I
responded by saying something like "Yeah, I'm the Pope, what can I
do for you?" After a few back and forths, I realized that the caller
was indeed a David Crockett and, in fact, a descendant of the Davy
Crockett who was a folk hero in Tennessee and the most famous
member of the losing team at the Alamo. The modern-day Crockett

was employed by IBM and was an elected member of the city council, where he'd become the primary force behind the greening of Chattanooga.

Once an industrial center of the Confederacy and later a major manufacturing center, the city had fallen on hard times. Industrialization has a bad habit of leaving ruin, pollution, and poverty in its wake while the folks who have made the money depart for higher ground, in this case to the bluffs above the city along the Tennessee River. Every year Chattanooga and Los Angeles fought over the bragging rights for the worst air quality in the United States. Crockett, however, did not think pollution, urban ruin, a necrotic downtown, and endemic poverty were the proper fate of the city and so was leading the effort to overcome the public inertia that accepts things as they are because people have forgotten what they once were and overlook what they might be. He called that afternoon to invite me to join the effort to improve Chattanooga. I signed on for a bit part.

David Crockett is a big man, six feet, six inches tall and three hundred pounds or more. He drove a beat-up sedan that is several sizes too small. It resembled a dumpster with headlights and a steering wheel—full of papers, books, cigarette butts, Coca-Cola cans, official documents, and whatever else happened to get dropped there. He is not a man who takes "no" for an answer and most of those who know him are happy to say "yes" anyway. His driving tours of Chattanooga are a visionary but hair-raising reminder of how short life can be. David has the driver's version of attention deficit disorder. He regards the paraphernalia of traffic safety like stop signs, stop lights, and posted speed limits merely as suggestions when he notices them at all. But he does notice the things that need fixing and improvement. "See that," he says, "we're gonna make that block a zero-waste manufacturing center . . . over there, that's a brownfield . . . gonna clean that up for downtown housing . . . that bridge is going to be a pedestrian bridge across . . . gotta be clean, green, and safe." And so it went.

In an annex to the city planning office and catty-corner from the historic Reed Hotel, Crockett held charrettes with some of the most notable pioneers of green urbanism in the United States and elsewhere. If you met there after hours a disco downstairs provided the rhythm, the second-hand smoke, and the buzz, while upstairs folks pored over giant maps of the city moving hypothetical things around, "brain storming," envisioning what could be made of and in this place. All the while Crockett choreographed the considerable ingenuity of the assembled talent and added color for good measure: a blue-blooded Tennessee politician he likened "to a turtle on a fence post . . . you know he didn't get there by hisself." After one particularly exciting hour of nonstop design enthusiasm he raised himself to his full stature and in a commanding voice said: "Boys, we're just like an east-facing hound dog at point . . ." Say what? No one got it until he added "on the median strip between the north-south lanes at rush hour." I'd never heard political reality described quite so vividly.

A quarter of a century later, the downtown core of Chattanooga, with its Tennessee Aquarium, pedestrian bridge across the Tennessee River, magnificent River walk, new housing, cleaned-up brownfield sites, parks, and more testifies to David Crockett's leadership and ability to make vision a reality. He won't admit it, but more than anyone else he was the sparkplug, orchestrator, and visionary behind the revitalization of the Chattanooga downtown. Along with a few others like Jaime Lerner, once the mayor of Curitiba, Brazil, Crockett helped to launch the green city movement and made it cool before it was the cool thing to do.[1]

The timing was important. Since its inception, the environmental movement had been busy fighting pollution, preserving wilderness and scenic rivers, and stopping all manner of bad things. It was in large measure an agrarian movement. Cities were mostly forgotten or treated as an afterthought. But Crockett knew that a decent human future would be mostly urban. Half of us now live in cities and the percentage keeps growing. In addition, cities generate 70 percent of the CO_2 emissions worldwide as well as the majority of other environmental impacts and policy innovations. In other words, cities matter.

Where most others saw only problems of ugliness, crime, pollution, and sprawl, David Crockett and the early pioneers of green urbanism saw possibilities and opportunities. Cities can be educational with aquaria, museums, nature centers, and universities. They can harbor a robust civic life that includes outdoor spaces for public debate, poetry readings, and art. They can foster conviviality in shaded, streetside cafes, as well as through street art, theaters, and music venues. They can include urban gardens and green roofs. They can mix the urban with rural, sometimes with a touch of wilderness. They can include bike lanes, walking trails, and light-rail transit, which provide mobility without automobile pollution and congestion. Cities can be clean, green, safe, fair, educational, and exciting incubators of human achievement and creativity. And with well-targeted policies and proper incentives, they can reduce a large fraction of the world's CO_2 emissions. The early green urbanists saw the possibilities that are rapidly becoming mainstream.

They also saw that making green cities required a different intellectual and policy framework. Cities are the most complicated and complex of human creations. Their pathologies—including crime, pollution, sprawl, and traffic congestion—have many causes, among them the fragmentation of functions by zone and the failure to reckon with the totality of the organism that must be fed, watered, sewered, informed, entertained, transported, and employed. In particular, its huge volume of waste—airborne, solid, and sludge—must be disposed of safely, and it must enable the movement of large numbers of people and massive amounts of food and materials every day. For all of their vibrancy and potential, all cities depend on long and vulnerable supply chains. Any failure in the systems supplying food, water, and electricity would cause chaos in a matter of hours and total breakdown in just a few days. There are other threats. Many coastal cities face the certainty of rising ocean levels and larger storms. Midcontinental cities will be exposed to prolonged droughts and larger and more frequent storms and tornadoes. Cities are also easy targets for those bearing any kind of grievance. They will always be vulnerable to hate groups, religious sects, terrorists, and the merely deranged. That is not a small thing in a world where the technological means to cause lethal havoc have been widely dispersed. Cities, in other words, are complex and vulnerable systems. Their future depends in large part on our understanding how complex systems work, how to make them more resilient in a multitude of ways, and how to design policy, law, and economic incentives to make them "clean, green, and safe." The study of the behavior of complex systems has a long history.[2]

The postwar decades between 1950 and 1980 were the grand era for systems theory. Building on advances in communications, operations research, and cybernetics from World War II, Kenneth Boulding, James G. Miller, Ludwig von Bertalanffy, C. West Churchman, Herbert A. Simon, Erwin Laszlo, Jay Forester, Dennis and Donella Meadows, Peter Senge, and others wrote persuasively about the power of systems analysis. The benefits were said to be many. Systems thinking would enable us to perceive the patterns that connect otherwise disparate things and to detect the counterintuitive logic underlying an often deceptive reality, thereby creating more coherent and effective analysis, plans, and policy.[3]

The actual benefits of systems theory, however, were mostly evident in computers and communications technologies, which in turn had mostly been based on advances in information theory and cybernetics during World War II. Elsewhere, business confidently marched on unperturbed. Despite the inherent logic of systems thinking, governments—as well as corporations, foundations, universities, and nonprofit organizations—still work mostly by breaking issues and problems into their parts and dealing with each in isolation. Separate agencies, departments, and organizations specialize in energy, land, food, air, water, wildlife, economy, finance, building regulations, urban policy, technology, health, and transportation, as if each were unrelated to the others. Consequently the right hand and left hand seldom know—or care—what the other is doing. The results, not surprisingly, are often counterproductive, overly expensive, risky, sometimes disastrous, and often ironic. Systems modeling, for example, has helped us to understand the causes of rapid climate change, whereas systemic failures in government, policy-making, and the economy have heretofore crippled our ability to do much about it. Systems theory, in short, has yet to have its Copernican moment and the reasons are ironically embedded in the scientific revolution.

Reducing wholes to parts, that is, "reductionism," lies at the core of the scientific worldview that we have inherited from Galileo, Bacon, and Descartes, among others. Reductionism works scientific, technological, and economic miracles. But as we gain in power, wealth, speed, convenience, control over nature, and self-confidence, we pay a considerable price, one that Faust (Marlowe's, not Goethe's) would have recognized. Like Faust, we have been short-termers, discounting long-term costs and risks that can only be seen from a systems perspective. The results have been dumbfounding. In record time we have shredded whole ecosystems, acidified the oceans, killed off thousands of species, squandered topsoil, leveled forests, and changed the chemistry of the atmosphere. "We are," in Edward Hoagland's words, "still part-chimpanzee with double degrees in trial and error." In the real world, as discussed earlier, things bite back, and there are tipping points, surprises, emergent properties, step-level changes, time delays—as well as unpredictable and catastrophic "black swan" events with long-lasting global effects. To anticipate and avoid such things requires a mindset capable of seeing connections, patterns, and systems, as well as a sightline far beyond the quarterly balance sheet or the next election. Wisdom begins with the awareness that we live amid complexities that we can never fully comprehend, let alone control. But caution was no part of the bulletproof exuberance written into our notions of progress or our foundational documents.[4]

The U.S. Constitution, conceived by men who were greatly influenced by the Enlightenment, ignorant of ecology, and fearful of excessive authority, gives no "clear, unambiguous textual foundation for federal environmental protection law." In other words, our manner of governing is often ecologically destructive because it does not deal holistically with whole systems. The 1970 National Environmental Policy Act aimed to remedy such shortcomings. It required all federal agencies to "utilize a systematic, interdisciplinary approach which will insure the integrated use of the natural and social sciences and the environmental design arts in planning and in decision-making." The act called for systems planning, but beyond the requirement for environmental impact statements for federally funded projects, it had no teeth. With few exceptions, things went on much as before.[5]

The upshot is that despite a great deal of talk about systems, we continue to administer, organize, analyze, manage, and govern complex ecological systems as if they were a collection of isolated parts and not an indissoluble union of energy, water, soils, microbes, land, forests, biota, and air. The concept of sustainability implicitly recognizes feedback and interdependencies among ecological, economic, social, and economic systems, hence a systems approach to environmental management. Alas, the reality is otherwise: the issues comprising sustainability have been fragmented so that isolated parts are no longer seen as one system.

Much of what we have learned about managing real systems began in agriculture, notably with the work of horticulturist Liberty Hyde Bailey, agronomist Albert Howard, forester Aldo Leopold, agroecologists Miguel Altieri and Stephen Gliessman, plant geneticist Wes Jackson, range management expert Alan Savory, and ecologically savvy farmers like Joel Salatin. The most important lesson from their collective work is that the land is a community of interrelated parts—soils, hydrology, biota, wildlife, plants, animals, and people. If sustainability is the goal, the land can neither be managed as a factory nor the profit it generates measured by its short-term yield. Managed as organism, the land limits the scale and kind of farming and forestry practices and eventually disappoints all expectations that exceed its carrying capacity. Good land stewardship requires patience, dependable long-term memory, wide margins otherwise called precaution, and, as Wendell Berry reminds us, affection. Its true profitability can be no greater than the rate at which sunshine can be turned into plant material and animal flesh without diminishing future productivity. In strict terms, a sustainable farm is one maintained in balance by natural inputs of sunlight, water, plant decay, animal manures, and by the observant and competent ecological intelligence of both the farmer and members of the surrounding culture. Like the natural systems it mimics, a sustainable farm is always a polyculture and depends on the synergies among its various components from soil microbes to animals. Industrial agriculture, in contrast, is subsidized by fossil fuels, imported fertility, chemical pest management, and borrowed capital. It is a part of the extractive economy that mines soils, minerals, genes, and people alike. And it is typically a monoculture aimed to make a profit in the short-term. The difference between industrial and ecological farming puts them at opposite ends of a continuum that defines resilience.

Ecologically designed buildings are another source of practical instruction about systems. Until the advent of the green building movement, the process occurred serially: architects did the basic design and passed the blueprints to the engineers to heat, cool, light, and plumb. The engineers, in turn, handed it off to the landscapers to make it look like it belonged where the happenstance of real estate prices and often bad planning had dropped it. The incentives, financial, legal, and reputational, demanded that the thing be overheated, overcooled, and overbuilt, which made it overly expensive. Much profit was made in excessive redundancy: it was as if a chairmaker made chairs with eight legs, because she was paid for each extra leg.

Early ecological designers such as Sim van der Ryn, Bob Berkebile, Bill McDonough, Pliny Fisk, and the U.S. Green Building Council pioneered a different approach to design that optimized the whole building as a system, not its separate components. A tighter, better-insulated building shell, for example, meant that HVAC systems could be downsized, reducing both the initial costs and long-term operating costs while improving human comfort. Similarly, creative daylighting improved aesthetics and occupant productivity while reducing lighting bills and, again, long-term costs. But the largest benefit of biophilic design lies in the very human fact that we are happier, healthier, and more productive in places carefully calibrated to our five senses.[6]

There are other areas of applied systems knowledge, but none so easy to grasp or as cogently instructive about the ways we might improve management of other systems. Yet these other systems each bring special challenges. Farming, for example, requires patience appropriate to the growing season and the long cycles that govern fertility. We can manipulate some of the variables inherent to farming, but the larger patterns of soils, hydrology, biota, wildlife, weather, and so forth have seasons and cycles to which we are strangers and interlopers. To the extent that we can manage them at all, prudence would have us leave wide margins to accommodate our ignorance and other shortcomings. Buildings, or what is awkwardly called "the built environment," by contrast, are a human creation. The designers are privy to their mysteries and inner workings in ways that we cannot know the natural systems of farms. Even so, builders are often surprised by the unanticipated behavior of mechanical systems, by design flaws and structural failures, and by the influence of human behavior on what were supposed to be well-conceived structures.

The human body is a third source of systems knowledge. Walter Cannon in his 1932 book *The Wisdom of the Body*, for example, introduced the notion of "homeostasis" as a way to explain how the "extraordinarily unstable material" of our bodies, in "free exchange with the outer world," miraculously persists through many decades. Yale professor and physician Sherwin Nuland, in a 1977 book with the same title, described the process in these words:

> Always on the alert for the omnipresent dangers without or within, ceaselessly sending mutually recognizable signals throughout its immensity of tissues, fluids, and cells, the animal body is a dynamism of responsible consistency. By untold trillions of energy-driven agencies of correctives, inappropriate alterations are balanced and changes are either accommodated or set right—all in the interest of that equilibrating steadiness that is the necessary condition of the order and harmony of complex living systems . . . Its capacity to communicate

within itself and with its external environment is the basis of an animal's viability in the face of the many unremitting forces that never cease to threaten its existence.[7]

The idea of the body as a complex system might have led to a systems view of medicine and healing, bridging the gap between Western and Eastern medicine. But the practice of Western medicine by that time was thoroughly reductionistic and immune to instruction from other cultures. Immersed in the Western science, physicians still tend to diagnose illnesses as if they had no deeper causes, heal diseases in isolation as if the body were a broken machine, and prescribe medications as if their effects did not ricochet throughout the body. The result is that solutions often become the source of new problems and the start of ever more vicious cycles.[8]

The same, however, could be said of most, if not all, fields of business, economy, public policy, and technology. For people trained to think in silos, working across boundaries does not come easily. But its essence is nonetheless straightforward: harmonious integration among the various components evident in ecological, social, and human health. So what can systems thinking teach us about how to better manage urban regions? In addressing the question of "what kind of problem a city is," Jane Jacobs wrote:

> Cities happen to be problems in organized complexity . . . present[ing] "situations in which a half-dozen or even several dozen quantities are all varying simultaneously and in subtly interconnected ways." Cities, again like the life sciences, do not exhibit *one* problem in organized complexity, which if understood explains all. They can be analyzed into many such problems or segments which as in the case of the life sciences, are also related with one another. The variables are many, but they are not helter-skelter; they are "interrelated into an organic whole."[9]

The challenge then is to transition from organized, urban systems built on an industrial model and designed for an autocentric growth economy into adaptive and resilient learning organizations in a rapidly changing world. Urban governments are being stressed in a world with more people, more stuff, and higher expectations, all moving faster. In Peter Senge's words, "Humankind has the capacity to create far more information than anyone can absorb, to foster far greater interdependency than anyone can manage, and to accelerate change far faster than anyone's ability to keep pace." The managers of urban systems require the capacity to shift "from seeing parts to seeing wholes, from seeing people as helpless reactors to seeing them as active participants in shaping their reality, from reacting to the present to creating the future." All of this is easier said than done. It will come as no surprise to city officials that city-regions, as Donella Meadows once wrote, are "self-organizing, nonlinear, feedback systems [and] are inherently unpredictable [so] . . . we can never fully understand our world, not in the way our reductionist science has led us to expect."[10]

The challenge to governance at all scales is to continually harmonize two kinds of nonlinear systems: (1) social and economic systems, which include laws, regulations, taxation, policies, elections, and markets and (2) ecological systems. These systems work on different time scales and by different processes as parts of a whole called the ecosphere. But they are not equal. Human contrivances—economy, technologies, politics, and social behavior—ultimately must conform to biophysical realities or they will disintegrate. We designed the systems by which we are governed and provisioned and we can redesign them, but only, in Meadows's words, if the people who manage them "pay close attention, participate flat out, and respond to feedback."[11]

A systems perspective to urban governance is a lens through which we might see more clearly through the fog of change and possibly better manage the complex cause-and-effect relationships between social and ecological phenomena. It would help make up for our chronic inability to foresee the consequences of our behavior. Knowledge of systems structures and operating rules may help improve our resilience in a rapidly warming world punctuated by black swan events and perhaps allow us to anticipate, rather than be surprised by, counterintuitive outcomes. The application of systems analysis is no panacea, but it does offer good opportunities for improving urban governance.

For one thing, systems analysis can help governments facing an overwhelming cacophony of raw data organize information so as to distinguish the ecological signals from the noise. A city is a complex and confusing array of inputs and outputs: fuels, food, materials, water, and so forth enter and carbon dioxide, wastewater, waste heat, pollutants, refuse, and all manner of other things exit. Were a city placed under an imaginary glass dome with the inputs and outputs entering and leaving through clearly marked pipes, we would understand these entropic flows and their interactions more directly. Short of that, it is possible to better understand a city through models that show ecological transactions as diligently as any accountant tracking the flow of money. Models of the city as a system of ecological inputs and outputs, then, are a useful tool for integrating disparate and confusing data into its larger ecological context in order to improve decision-making across sectors, departments, and agencies.

In addition, the data necessary to understand resource flows and the larger ecological context of a city can be deployed creatively to educate its citizenry to understand the relationships between their collective behavior and the environmental and economic prospects. The use of flat-screen displays (dashboards) in buildings, city kiosks, sports arenas, libraries, hotels, and schools in order to track and display data on resource flows, carbon emissions, investment, land-use patterns, ownership, public attitudes, and the interactions of all of these can be a powerful tool for educating citizens about feedback loops and delays between action and results, as well as for improving their understanding of complex issues. The result could be a widely accessible and cost-effective education in the basic dynamics of biophysical, social, and economic interactions.

Systems analysis can help to improve planning and forecasting. Elected leaders in many rust-belt cities like Detroit, Cleveland, and Youngstown assumed that the good times would never end and were caught flat-footed when they did end. The use of models that clarify assumptions, identify feedback loops, and monitor system behavior and ecological conditions can help decisionmakers to better anticipate change and to plan, tax, budget, and craft smarter policies. Looking ahead, cities in a rapidly warming world must prepare for larger storms, longer droughts, supply disruptions, and economic turmoil. These near certainties, in turn, ought to affect decisions about zoning, land-use, location and types of infrastructure, building codes, food supplies, economic development, taxation, and emergency preparedness.

The tools of systems analysis can also help to improve the quality of urban decision-making. To get a driver's license, for example, one must take a course and pass a test. But for officials charged with managing the public business, we require virtually no evidence of any basic understanding of how the world works as a physical system and the processes that govern the interactions of social and natural systems. As noted earlier, we would be properly intolerant of public officials who were unable to read or count, but ecological illiteracy—a more serious problem—causes no consternation whatsoever. As part of their routine orientation to city government, officials, elected and appointed, ought to be required to pass a basic test in ecology and systems dynamics. However carried out, the goals would be (1) to raise the effectiveness of decision-making by increasing leaders' awareness of how urban regions, as social and economic systems, interact with natural systems, and (2) to equip leaders with better tools of analysis and foresight with which to manage public business.

Systems analysis can improve organizational behavior as well. The capacity to respond to feedback is inhibited by many factors. It can be blocked when fear, group-think, or complacency paralyze decision-making. Rather than suppressing dissent, systems analysis can help to clarify the unexamined differences in beliefs embedded in competing paradigms and mental models. David Cooperrider and Peter Senge have developed techniques to facilitate systems thinking in order to build organizational community around common visions. Their goal is to enable members of organizations to see themselves as players in an enterprise that makes decisions involving feedbacks, step-level changes, emergent properties, stocks, and flows that advance the awareness of agency in causing one outcome or another.

Further, systems thinking can lead to greater realism and precautionary public policies for the simple reason that the behavior of systems is inherently unpredictable, that is to say, nonlinear. It follows that policies of all kinds should be designed with wide margins, hedged bets, and redundancy. Every intended solution should solve more than one problem while causing no new ones. The goal, in short, is to build smarter and more adaptable institutions and organizations that are capable of learning, foresight, and intelligent agency, and that are "robust to error," at the intersection of human action and biophysical realities.

From a systems perspective there are no such things as "side effects," only the logical outcomes derived from the rules and behavior of the system. Climate change; ozone holes; cancer clusters; and Texas-sized, mile-deep gyres of trash floating in the middle of the Pacific Ocean are not side effects of economic growth, but the predictable outcomes of a system designed to grow at all costs. Likewise, from a systems perspective there are few accidents, only the institutionalized lack of foresight, which is a flaw in the way a particular system is organized. The point, in Senge's words, is that "everyone shares responsibility for problems generated by a system."[12]

Finally, systems analysis of a city's building stock, infrastructure, patterns of energy use, and utility data, in conjunction with smart policies and financial incentives, can be used to reduce CO_2 emissions without federal action. City-regions, as Jane Jacobs thought, are drivers in the national economy, not passive actors. The same may be true for establishing climate policies. Political scientist Benjamin Barber argues that the "immediacy and the local character of city politics" means that cities are better positioned to act quickly and effectively than any other level of government. Mayors and city councils control building codes, administer zoning regulations, and have considerable leverage on CO_2 emissions due to their buying and investing power. If they were to join larger organizations such as the C-40 cities, the Compact of Mayors, and the Carbon Neutral Cities Alliance, there is the possibility that cities can be a significant part of the effort to hold the temperature increase to the two degrees Celsius target and to drive federal policy. The same is more obvious at the state and regional levels. California, led by Governor Jerry Brown, has implemented sweeping changes in energy policies and urban development aimed to cut CO_2 emissions in half by 2030. In each case the essential policy and financial tools are those of systems analysis that integrate data from otherwise dispersed fields and extend understanding of consequences farther in time.[13]

Cities and states, however, can't do it on their own. They will need significant help. Many will be flooded with refugees both international and domestic. Some will be literally flooded by rising sea levels. All will need financial help in meeting the additional costs of adapting to those changes for which adaptation is possible. They will need additional help with the security necessary to keep order. They will need help in responding to a rising number of climate-driven emergencies. They will need major assistance to maintain the electric grid and supply chains. And so forth. In short, we have no choice but to become a nation that can function effectively and with foresight at all levels from rural areas to mega-cities, that is to say as a coherent, resilient, and democratic system at a continental scale.

The goal of systems analysis and organizational learning is not just to find a more clever way for cities and other organizations to do what they have been doing all along. It is rather a tool to help reexamine purposes and performance relative to complex and rapidly changing circumstances. As with any tool, its effectiveness depends on the skill and wisdom of the user. Systems analysis is not magic; it cannot tell us what to model or what's worth doing and what's not. It will not make the stupid and hardhearted wise or caring. It won't tell us anything that lies outside our paradigms, worldviews, or beyond the light of our particular campfire. It is, after all, only a tool and will do no more than what it is asked to do and no more than any culturally constrained and time-bound technique can ever do. We have to supply the compassion and good judgment, and care enough to want to know the consequences of our actions. Further, there is nothing new in systems thinking beyond the higher level of precision and analytical power inherent in sophisticated computer modeling. Earlier societies created complex ways to both foresee and restrain certain behaviors that could damage their collective prospects. The Amish achieve many of the same results by maintaining a coherent and sober, if restrictive, culture.[14]

In the end, systems analysis applied at the level of organizations, cities, and regional governance buys us time until, among other things, national governments catch up. At any level, it is only a tool to clarify the consequences of our actions, identify our options, and extend our foresight a bit. Yet those are not small gains.

Oberlin commenst this war. Oberlin wuz the prime cause
uv all the trubble.

—Petroleum V. Nasby, 1873

Oberlin, Ohio, a city of about 10,000, is located fourteen miles
from Lake Erie, thirty-five miles from Cleveland, and as a sober crow
would fly it, eighty-four miles from Detroit. Situated on the till left
when the last glaciers retreated twelve thousand years ago, the city
today lies in the geographic center of the rusty industrial heart of
America. The town was founded in 1832 by do-gooders who drove
out the natives, bears, and wolves, and tried to improve the tougher
wildlife—the hard-drinking frontiersmen who preferred the back-
woods to the comforts of civilization. The city grew around the college, which had been named for John Frederic Oberlin, an Alsatian
pastor famous for saving souls and improving the local infrastructure of roads, bridges, schools, and hospitals. Early in its history
Oberlin College accepted African Americans and women as full students and became a busy stopover along the Underground Railroad.
It has been said that this is the "town that started the Civil War" by
rescuing a runaway slave by the name of John Price from Kentucky
bounty hunters in 1858. Ever since, Oberlinians have prided themselves on being a step ahead, marching to the superior rhythm of a
more progressive drummer.[1]

The city of Oberlin is formed around a thirteen-acre square named
for the Tappan brothers, New York City businessmen and abolition-
ists who supported the college in its early years. Oberlin once had
six downtown grocery stores, two drug stores, an urban trolley system, and rail connections to anywhere in the United States. While a
student at Oberlin in 1886 Charles Martin Hall discovered how to

separate aluminum from bauxite, which eventually led to the founding of the Alcoa Corporation. The downtown retains a quaint nineteenth-century charm, but merchants selling anything other than beer, pizza, or coffee struggle to make a go of it. Oberlin College and its distinguished conservatory of music anchor the economy, and it is home to the largest air-traffic control center in the United States and a scattering of other businesses, but the total does not add up to a robust economy. North of Oberlin, two rust-belt cities have been drained by the neglect and disinvestment characteristic of American urban policies after World War II. If a ragtag bunch of foreign terrorists had done one thousandth as much damage, our patriotic vengeance would have known no limits. Alas we did it to ourselves, so we applauded the footloose brigands who left ruined towns and shattered communities from Flint and Detroit to Youngstown and beyond. Driving south from Oberlin down state highway 58 you will see mostly farmland and scattered woods, deer, raccoons, Amish farmers, and conservatives.

A few years from now, Oberlin, like every other place on Earth, will face mounting challenges from rapid climate change. If our current policies continue unimpeded, the planet could be a lot hotter by the year 2100. Between now and then many things will come undone starting with our water and food supplies, but eventually—perhaps sooner than later—entire economies and political systems. Nearly everything on Earth behaves or works differently at higher temperatures. Ecologies collapse, forests burn, metals expand, concrete runways buckle, rivers dry up, dust bowls grow, and people stressed by heat come undone more easily. No place will be spared. Oberlin, too, will experience its share of hotter and less predictable weather, larger storms, bigger floods, longer and more severe droughts, changing seasons, shifting ecologies, and stressed-out people. I'll wager that before long an influx of folks with southern accents will be migrating out of harm's way, as they say, toward Great Lakes water and a slightly more benign climate. They won't be so much opportunity seekers as refugees, much like the Okies fleeing the 1930s Dust Bowl. Given the vacuum of leadership on climate and energy policy in Washington and in states like Ohio, what can be done? In the face of reality we will need practical visions.

Soon after I moved to Oberlin in 1990 a battle-scarred veteran of local politics told me, "Son, you gotta understand one thing about this town . . . If you put all Oberlinians in a burning building they could not agree on how to get out. Just accept it." I never did and no one else should either. Authentic vision and foresight are essential to a decent common future. But what kinds of visions are appropriate to the twenty-first century, rapid climate change, and a population trending toward eleven billion? At any scale, from small communities to the global commons, how do worthy visions come to be? Does it matter whether they come from the market or from the political process? What is the role of leadership versus public participation? Or inspiration versus analysis? How do visions become ingrained in the behavior of organizations and in the lives of people? How can they be modified and adapted to continual change?

Whatever the answers, our history and cultural DNA have conspired to make collective visions and planning more difficult than they should be. As Bill Clinton puts it, "We share 99.5 percent of our genes but spend 99.5 percent of our time arguing about the .5 percent of our differences." We are a contentious "argument society," in Deborah Tannen's words, burrowed down in our favorite blogs and websites and selecting our news from versions form-fitted to our particular biases. Echoing Thomas Jefferson's streak of anti-government obstinacy, common vision or common anything comes hard for us these days. Maybe Madison and de Tocqueville saw this coming, but I doubt that they thought it would be like this at either the national or community scale. "The vision thing," in its collective form, seems to be difficult for us, a people who exhausted ourselves corralling the natives, cutting down entire forests, ripping up prairies, damming rivers, wiping out untold numbers of species, corporatizing our economy, paving over an area larger than the state of Kentucky, lopping the tops off Appalachian mountains for cheap coal, drilling several million holes in the ground to extract oil, winning a couple of World Wars and one Cold War, watching a lot of crummy TV, selling truckloads of junk to each other, building a global empire, tweeting and blogging nonsense nonstop, and becoming the richest, most rotund, and most self-congratulatory people on Earth. But the collapse of the Soviet Union in 1990 left us adrift and despondent as the sole superpower with no dependably loathsome enemies in sight. And so we began beating up on each other in phony culture wars, government shutdowns, and increasingly vicious and sterile political battles. Until, that is, Osama bin Laden showed up in the nick of time and created the rationale for steroidalized (a new word) superpatriots to create the twenty-first-century version of Orwell's *1984* dystopic vision: the corporate-garrison-surveillance-entertainment state. In the meantime far more serious threats were growing.

Five reports from the Intergovernmental Panel between 1990 and 2013 and an Everest-scale mountain of other scientific and anecdotal evidence give us reason neither for complacency nor cheap optimism. The longevity of CO_2 in the atmosphere "may change the climate for millennia to come," locking us into a future of worsening disasters. The possibility of creating the future we might otherwise have wanted is long gone. Our best hope is to own up to the future we've made and cannot evade, get busy to forestall the worst that could happen, and lay the foundation for a better world on a farther horizon.[2]

But how do we grow practical visions at any scale appropriate to the conditions of this new and less stable geologic era, the "Anthropocene"? Without vision, people perish, which is to say that if we intend to be around for the long term, foresight is not optional. The first issue, however, is whether enough people are willing to see the truth. But sight, the willingness to see, is partly a matter of choice. Psychologist Oliver Sacks once had a patient, blind from early childhood into middle age, who with medical help recovered most of his sight. Then an extraordinary thing happened: he decided that his sightless fantasy world was preferable to reality and chose to return to his former blindness. Sacks concluded that sight is at least partly optional. The will to see the world accurately, too, is a choice and many people prefer fantasy, ideology, and denial in some form or other to messy reality. Foresight—seeing into the future—is more difficult still. And like

Cassandra in Greek mythology, all of those who rain on other people's parties are subjected to ridicule, rejection, or worse, simply being ignored.

But from time to time practical visions can grow in a variety of ways. Sometimes they result from the "wisdom of crowds" and social changes. For example, between 2005 and 2010 consumers in eleven western nations saved $429 billion worth of oil through improved efficiency. Today, too, we create a dollar of GNP by expending only half the energy that we did for the same effect in 1973. The big drivers of these improvements are technological advances and the desire to save money, not changes in public policy. Looking to 2050, Amory Lovins believes that we could run a far larger economy without using coal, oil, or nuclear energy, and a third less natural gas. He doesn't say what good use we'd make of a larger economy, but that's a different matter.

Some believe that we can grow worthy visions from the fertile soil of self-interest manifest in markets. Economist Karl Polanyi's dissent, noted earlier, has been vindicated in the subsequent history of the minimally supervised market. Market visions were more of the desert mirage sort, and ended up giving us urban sprawl, decaying cities, highway congestion, crime in high and low places, mega-malls, and massive political and financial corruption—and far fewer coherent neighborhoods, vital downtowns, prosperous homegrown businesses, or good rail service, to say nothing of a robust democracy, integrity in high places, and basic fairness. The problem may be that the "free market" isn't really all that free.

Vision sometimes emerges from the free market of ideas. Harriet Beecher Stowe's book *Uncle Tom's Cabin* or Karl Marx's *Das Kapital* changed minds and behavior and shifted the course of history. Ernest Callenbach's 1975 book *Ecotopia* was widely read for decades and helped to shape the mindset that led to the Cascadian vision and culture of the Pacific Northwest. In conservative philosopher Richard Weaver's words, ideas do have consequences and sometimes, for better or worse, they change the course of history.

Common vision can also grow from organized community dialogue. The turnaround of Chattanooga described in Chapter 11 began with a series of community-wide planning meetings during which residents were asked to respond to visual preference surveys that provided the basis for subsequent plans for a river walk along the Tennessee River, restoration of brownfield areas, creation of parks and housing, a new Tennessee Aquarium, and downtown commercial and residential development. The urban greening movement pioneered by David Crockett in Chattanooga, Jaime Lerner in Curitiba, Scott Bernstein in Chicago, and others is similarly thriving in large cities like New York, Philadelphia, Chicago, Cleveland, Los Angeles, Portland, and Seattle. Douglas Farr, Katherine Gejewski, Sadhu Johnston, Jenita McGowan, Julia Parzen, and others have used urban sustainability offices as effective platforms for growing vision and building the capacity to act on it. The EcoDistrict and Transition Town movements are similarly enabling neighborhoods and smaller cities to reduce carbon emissions, develop community gardens, develop local enterprise, and rebuild self-reliant communities.

James Fishkin and Bruce Ackerman, as noted earlier, have proposed grassroots efforts to revitalize democratic dialogue through "deliberation days" in which citizens come together on the same day in different cities in order to discuss important issues in an organized, thoughtful way. With proper facilitation and accurate information, people can talk through tough problems, narrow their disagreements, and sometimes reach a common vision that transcends right and left.

In sum, vision can come from market trends, social movements, or powerful ideas presented at the right time. It can come from astute and committed leadership. It can come from grassroots deliberation and initiatives or from top-down laws and regulations. It comes in all sizes and shapes and emerges from different places, cultures, and times. But grand visions are not always benign. Exclusionary movements can flourish when the fabric of tolerance and decency is frayed by prolonged adversity. And that fact makes our efforts to build decent and sustainable communities everywhere that much more urgent. If past history is any guide, the trouble begins early on, as pressures mount on family budgets, businesses, community services, and everyone's patience. Tolerance, decency, and truth are early victims.

Whatever their source, visions must be transformed into the routine behavior of urban governments, institutions, and organizations in order to be effective. Doing the right thing must become the easy and normal course of action and so requires recalibrating prices, taxation, regulation, financial management, public spending, information flows, and incentives to ecological realities. This recalibration, in turn, requires a supportive civic culture capable of anticipation, learning, and smart collective action.

In Oberlin, I live, like everyone else, in a world warming rapidly from too much fire. Against that hard reality, the college community has aimed to create a vision of a city and inclusive local economy powered by sunlight and efficiency. To have any chance of success in a hotter, stormier, steamier, buggier, and drought-stricken world we must come together on higher ground. Concern about these issues led in 2009 to the creation of the Oberlin Project, which is a joint effort of the city and college to develop a model of "full spectrum sustainability" working across sectors of energy, education, agriculture, policy, finance, and urban development. In plain English, it is an attempt to avoid the usual bureaucratic fragmentation and to "connect the dots" between the various parts of sustainability, using systems-based knowledge and extending the time horizon by which we judge our successes and failures. In practical terms, it means having lunch with many different kinds of people and attending lots of meetings to bridge the chasms that divide us by issue-areas, functions, class, and political affiliation. Since systemic failures have led us to the present crisis, we assume that emerging safely from it will require systems-level responses, smarter policies, and alert citizens acting with foresight and civic acumen.

Specifically, the Oberlin Project is an integrated response to the many challenges posed by climate change focused around seven practical goals:

1. Develop a thirteen-acre district in the downtown (the Green Arts District) at the U.S. Green Building Council's "platinum"

level as the main driver for community economic revitalization. Development projects in the district will include restoration of a famous art museum (completed); restoration and expansion of a performing arts center; and construction of a hotel, conference center, and business complex completed in 2016. We intend that the downtown renaissance will create local employment, grow the economy, and establish a new benchmark for community-scale green development.

2. Create new businesses in housing, energy efficiency, and solar deployment. In the transition to carbon-neutral sustainability, we propose to create and expand locally owned businesses and promote widespread ownership that spreads wealth throughout the city and thereby increases economic resilience.

3. Shift the city and college to renewable energy sources, radically improve efficiency, eliminate our carbon emissions, and improve the local economy. City residents and businesses presently spend about $15 million each year on electricity and natural gas—twice as much as we would need to spend if we were as efficient as is now economically advantageous and technologically feasible. We aim to reduce energy use by improving efficiency (and so saving millions of dollars), building a local renewable-energy economy that creates jobs and ownership, and growing the local economy while buffering Oberlin from rising energy prices and sudden cost increases.

4. Establish a robust local foods economy that provides a growing percentage of our food need and supports local farmers. Presently, only a small fraction of what Oberlin residents eat is grown in northeast Ohio, which means that money is unnecessarily flowing out of the community. We propose to expand the market for locally grown foods, which will improve the local farm economy and create new employment opportunities in farming (including summer jobs for teens) and food processing, all while improving the taste and nutritional quality of food we eat.

5. Expand educational collaborations that enhance teaching and learning about the challenges and opportunities of sustainability

among Oberlin College, the Oberlin schools, a nearby vocational school, and Lorain County Community College.

6. Broaden and deepen the local conversation on sustainability to include all of the humanities, all of the arts, the sciences, and the social sciences.

7. Collaborate with similar projects and communities across the United States.

With the onset of rapid climate change, our choice is not whether we do such things, but whether we do them as an integrated, well-thought-out system in which the parts reinforce the resilience and prosperity of the entire community or instead as a series of disjointed, one-off, overly expensive ad hoc responses to external crises, supply interruptions, and volatile prices.

The Oberlin Project is designed to be catalytic, not permanent. In other words, we intend to do our work in a few years to make sustainability the default and get out of the way. To that end we have organized public committees on local food, energy, housing, economic development, and education. To avoid creating yet more silos, we seek to foster collaboration with other organizations in the city to find synergies where $2 + 2 = 22$, not just 4. We propose, in other words, to give practical meaning to the idea of systems in the day-to-day affairs of the city, the college, and the local economy.

Early on we circulated Donella Meadows's paper "Places to Intervene in the System" to begin a dialogue about where, when, and how to effectively intervene in the systems of the city and the college. But the fact is that there is no one place to intervene that works in every city, in all circumstances, on every issue, all of the time. Strategies for change accordingly must be flexible, calibrated to locality, situation, culture, and institutional contexts.[3]

We began with the practical and doable objective of weaning ourselves from all fossil fuels. At this writing (May 2016), we have eliminated approximately 90 percent of our CO_2 emissions from the city-owned utility. We are a member of the the Clinton Foundation's C40 cities. We have been one of seventeen cities selected as "Climate Champion Cities" by the White House. We've deployed a 2.27 megawatt solar array on eleven acres and are planning to add another 2.75 megawatts in the next few years as a community-owned solar cooperative.

But we have a long way to go to become sustainable. We set a goal of growing as much as 70 percent of our food locally. This is a tall order and we will need to relearn a great deal about natural systems agriculture and management of farms and gardens. We've started a business incubator to nurture local entrepreneurs and build a sustainable economy from local talent. We have begun a food hub as a broker between farmers and the local food market. Even more urgently, however, we need to grow a generation of leaders from the presently unemployed, underutilized, under educated, and drifting. We need their energy, their smarts, and their hearts hitched to a future that they help to dream and build. We need citizens who know how the world works as a physical system and who understand how, when, and where to intervene in complex systems to cause the right kinds of change at the right time. We need peacemakers, dreamers, doers, and wise elders. We need people who make charity and civility the norm. We need more parks, farmers' markets, bike trails, baseball teams, book groups, poetry readings, good coffee, conviviality, practical competence, and communities where the word "neighbor" is a verb, not a noun. We need people who know and love this place and see it whole and see it for what it can be.

The challenges of sustainability and improving resilience in the face of rapid climate change require expanding the typical range of services. Most larger cities now have offices of sustainability, climate action plans, and plans for smart-growth and eco-districts. But they will have to do a great deal more. Effective responses to rapid climate destabilization will require carefully designed policies that improve energy efficiency, beginning with the large energy users in the commercial building and manufacturing sectors. They will require public and private incentives that encourage rapid deployment of renewable energy and that take full advantage of a growing array of ongoing technological advances including those in energy distribution ("smart grid" innovations). They will require better information beginning with accurate and publicly accessible models of material flows, carbon emissions, financial data, and public attitudes. They will require restoring local, urban, and regional food systems as farming elsewhere is stressed by heat and drought. Effective responses to changing patterns of rainfall that range between drought and massive storms will require rebuilding infrastructure and water storage systems, upgrading building codes to accommodate higher winds and temperatures and larger floods, and improving emergency response capabilities. Communities must also adopt policies and laws that promote sustainable economic development and require full-cost prices.[4]

In short, we will have to redesign a great deal of the physical infrastructure that worked in the brief age of cheap fossil fuels, along with the policies, tax codes, subsidies, and other incentives that made fossil fuels profitable for a few while making changes hard for everyone else. The challenges are daunting and long-term, but the technological know-how, design capabilities, architectural skills, urban-planning capabilities, engineering, and "idea capital" necessary to make the transition already exist. Even so, we have yet to build a citizenry that understands the scale and duration of the challenge and what it will require of them. For full-spectrum sustainability is neither a more clever way of doing the same old things, nor does it involve simply tinkering with the coefficients of change. Instead it demands a transformation of the structure of the systems that have rendered our future precarious. Full-spectrum sustainability requires that we learn to see the world—and ourselves—whole and apply the intelligence, foresight, generosity of spirit, and civic competence to avoid unsolvable dilemmas and fix problems before they become full-blown crises.[5]

How far have we—the college, city, and community—come? In the first seven years the milestones for the Oberlin Project include:

- renovation of a nationally prominent art museum in the Green Arts District at the USGBC Gold level;

- completion of a $17 million privately financed downtown housing and commercial development project, also at the USGBC Gold level;
- renovation of a historic downtown theater;
- selection as one of eighteen Clinton Climate Positive projects worldwide (now part of the C40 cities);
- selection as one of seventeen "Climate Champion" cities by the White House;
- deployment of a 2.27 megawatt photovoltaic system;
- a municipal electric supply that is more than 90 percent carbon-free;
- organization of community-based teams focused on energy, economic renewal, education, and food/agriculture;
- completion of a $1.1 million study, funded by the U.S. Department of Agriculture, on the regional transition to energy efficiency and renewable energy;
- adoption of a citywide climate action plan;
- development of a college plan for climate neutrality by 2025; and
- completion of a platinum-rated hotel and conference center powered entirely by solar energy.

The Oberlin Project is many things depending on one's vantage point, including:

- a city-scale experiment in the art and science of integrated solutions;
- an educational experiment that engages students in the design and development of a model of integrated sustainability that pertains to virtually every department and discipline;
- a model of homegrown, post-cheap-fossil-fuel economic revitalization;
- an improved foods system that provides opportunities for good work in a healthier community with more physical activity, wholesome food, and cleaner air and water;
- a community-scale model of resilience that reduces present and future residents' vulnerability to outside disruption,

whether from malice, technological accidents, or rapid climate change;[6] and

- a source of local pride in a small community with a long history of standing up when it counts.

But most important, it is an exercise in applied hope based on a commitment to make the world more fair and decent while preserving a beautiful and livable Earth. And if we don't stand for such things, what do we stand for?

In a larger perspective of time and geography, Oberlin is a very small drop in a very large sea. The Oberlin Project is only a bench-lab-scale experiment, small enough to be agile and instructive but large enough to be significant in the wider scheme of things. We who live and work in Oberlin have a unique combination of history and institutional capabilities that predispose us to give priority to issues of justice, to harness the power of the arts and music in the cause of human betterment, and to think across the conventional divisions of departments and disciplines. But every town, city, and region in the United States has its own unique blend of assets and possibilities. Imaginatively developed, they can singly and in combination affect issues beyond their borders, thereby becoming catalysts for larger change. This is the kind of revolution from the bottom that Lewis Mumford envisioned: a citizens' movement of transition towns, eco-districts, and green cities that may one day provoke an outbreak of sanity and better purposes in state capitals and even Washington, D.C.

Still larger possibilities can be organized at a regional scale, as Mumford and others proposed in the 1930s. For example, the cities that rim the western shore of Lake Erie—Flint, Detroit, Toledo, Cleveland, and Youngstown—were once the industrial heart of the U.S. economy. Now they represent our greatest economic and social challenges, as well as perhaps our greatest opportunity for economic renewal as a nation.[7]

The Lake Erie Crescent region is geographically coherent. It lies within a common watershed. It has a common history and identity.

It still has a skilled work force. It has a stable core of foundations, universities, corporations, and civic organizations forged in prosperity but tempered in adversity. And it is proximate to the Great Lakes, the largest collective body of fresh water in the world. It has, in other words, assets that will increase in value. The renaissance of the Lake Erie Crescent can play a prominent role in the transition toward a prosperous nation that is sustainable, resilient, and powered by renewable energy. Business, governments, and philanthropy all have a role to play. But colleges and universities can play a much larger role than they have until now by collaboratively redirecting their buying and investment to accelerate the deployment of solar energy and energy efficiency; resuscitate a robust regional food system to displace imports; and revitalize urban cores, through smart growth and by adopting advanced design standards for building and community design.

Buying from regional manufacturers, growers, and service providers is a strategy that Jane Jacobs has called "import substitution." It promotes economic development by taking advantage of economic multipliers that retain profits in local and regional economies. The advantages include reduced carbon emissions associated with long-distance transport and increased employment and human capital in the region. Purchasing within the region means increased resilience and a faster transition to a zero-carbon economy, one driven by smart urban growth, robust sustainable agriculture, and renewable energy technologies. And since it redirects monies already budgeted, it does not require massive new federal or state assistance. Further, it takes advantage of three rapidly developing trends: affordable renewable energy, local and sustainably grown foods, and green building.

The same logic applies as well to the strategic deployment of investments from university endowments to rebuild regional systems, downtown cores, and support local businesses, farmers, and service providers. Capital redeployed to help redevelop regions might avoid the pitfalls of politically mandated change. The key players in such an initiative would be college and university presidents and their chief financial officers and investment managers. To be successful the process would require (1) collaboration and information sharing about purchasing and investment opportunities in the region, (2) new rules for managing capital that recognize its multiple forms as natural capital, human resources, financial assets, culture, and ecosystem services, (3) strategies that would transition the system from one that demands high returns and lowest costs in the short term to one that builds long-term, stable prosperity and creates collateral benefits such as employment, lower crime rates, lower carbon emissions, cleaner air, and restored urban cores, and (4) means by which such collateral benefits can be monetized and rewarded.

Finally, regional collaboration led by anchor institutions with deep roots and long time horizons can be replicated in many places throughout the United States. In addition to colleges and universities, healthcare facilities, sports teams, public institutions like zoos and libraries, and military facilities could work to realign procurement and investment with basic human needs in the long emergency.

AFTERWORD

I am writing these words while on a family vacation on the outer banks of North Carolina with our two sons, their wives, and four grandchildren who range in age from six to sixteen. My thirteen-year-old granddaughter asked whether these beaches and dunes would soon be underwater. I said "maybe" and no more. But the fact is that their children, should they choose to have any, will not see this place as it exists now. It will have long before been drowned by rising seas. Nor will they ever know the hemlock forests that I knew as a boy roaming the hills of western Pennsylvania. Much of the natural beauty I've known will be lost to them. Maybe they won't notice, because we are an adaptable species and our perceptions and expectations shift downward as things worsen. They will experience other places, perhaps mostly indoors. Losses, whether of species, landscapes, or seascape—a world without—will become simply the new default. But whether we notice them or not, these losses matter because each diminishes our experiences, pleasures, and possibilities. Each loss is a steady diminution from the world that shaped our humanity and informed our senses. Each loss of a species is a loss of a companion on the long journey of evolution. Each place lost to an encroaching desert or rising waters is a loss of beauty, memory, and sustenance. Each turn of the ratchet downward signifies lost opportunities for human flourishing. What does one say to a grandchild?

NOTES

Prologue

1 Elizabeth Kolbert is skeptical: see her "Project Exodus," *New Yorker*, June 1, 2015, pp. 76–79. Diplomat and historian Louis Halle was not: refer to his "A Hopeful Future for Humankind," *Foreign Affairs* (Summer 1980).

2 Jack Miles, "Global Requiem: The Apocalyptic Moment in Religion, Science, and Art," *CrossCurrents* (January 2001).

3 For example, see biologist Carl McDaniel's perceptive account of the 1910 story of the British ship *Endurance* caught in the ice of Antarctica. Facing sure destruction, the ship's captain, Ernest Shackleton, led his beleaguered crew to safety crossing hundreds of miles of ice and open waters. Shackleton's extraordinary leadership under extreme duress drew from a combination of personal qualities of forbearance, cool judgment, brilliant psychology, as well as his mastery of the physics of ice, direction, ocean currents, and navigation. See Carl McDaniel, *At the Mercy of Nature* (Medina, OH: Sigel Press, 2014). Edward O. Wilson writes: "Scientists who might contribute to a more realistic worldview are especially disappointing . . . they are intellectual dwarves content to stay within the narrow specialties for which they were trained and are paid." Wilson, *The Meaning of Human Existence* (New York: Liveright, 2014), p. 178.

4 Brendon Larson, *Metaphors for Environmental Sustainability* (New Haven: Yale University Press, 2011) is an insightful and cautionary analysis of the use of metaphors.

Chapter 1. A Perfect Storm

Epigraphs. Norbert Weiner, *The Human Use of Human Beings* (1950; New York: Avon Books, 1967), p. 252; James Martin, *The Meaning of the 21st Century* (New York: Riverhead Books, 2006), p. 373.

1 George Perkins Marsh, *Man and Nature* (1864; Cambridge, MA: Harvard University Press, 1967), p. 36; Fairfield Osborn, *Our Plundered Planet* (Boston: Little Brown, 1948); William Vogt, *Road to Survival* (New York: William Sloane Associates, 1948).

2 William L. Thomas, Jr., ed., *Man's Role in Changing the Face of the Earth* (Chicago: University of Chicago Press, 1956); and the later update, B. L. Turner et al., *The Earth as Transformed by Human Action* (Cambridge, Eng.: Cambridge University Press, 1990); Sandra Steingraber, *Living Downstream* (Reading, MA: Addison-Wesley, 1997); Theo Colborn, Dianne Dumanoski, and John Peterson Myers, *Our Stolen Future* (New York: Dutton, 1996).

3 Eugene Odum, *Fundamentals of Ecology* (Philadelphia: Saunders, 1951); Donella Meadows et al., *The Limits to Growth* (New York: Universe Books, 1972); W. Steffen et al., *Global Change and the Earth System* (Berlin: Springer, 2004); W. Steffen et al., "Planetary Boundaries," *ScienceExpress*, January 15, 2015; Johan Rockström et al., "Planetary Boundaries: Exploring the Safe Operating Space for Humanity," *Ecology and Society* 14, no. 2 (2009); Johan Rockström and Mattias Klum, *Big World, Small Planet* (New Haven: Yale University Press, 2015); A. Barnosky et al., "Approaching a State Shift in Earth's Biosphere," *Nature* 486 (June 7, 2012): 52–58.

4 William Catton, *Overshoot: The Ecological Basis of Revolutionary Change* (Urbana: University of Illinois, 1980).

5 For perspective, see Will Steffen et al., "The Anthropocene: Conceptual and Historical Perspectives," *Philosophical Transactions of the Royal Society A* 369 (2011): 842–867.

6 Eric Schlosser, *Command and Control* (New York: Penguin Books, 2013); David Hoffman, *The Dead Hand* (New York: Doubleday, 2009); Ron Rosenbaum, *How the End Begins* (New York: Simon & Schuster, 2011); Associated Press, "Key Findings on Nuclear Force Troubles," *New York Times*, January 27, 2014; Helene Cooper, "Cheating Accusations among Officers Overseeing Nuclear Arms," *New York Times*, January 15, 2014; Philip Taubman, *The Partnership* (New York: Harper Perennial, 2012); Elaine Scarry, *Thermonuclear Monarchy* (New York: Norton, 2014); Robert S. McNamara, "Apocalypse Soon," *Foreign Policy* (May–June 2005): 29–35; Lee Butler, "Zero Tolerance," *Bulletin of the Atomic Scientists* 56, no. 1 (January–February 2008). But "massive weapons upgrades" presently under way undermine the zero tolerance goal. See John Mecklin, "Disarm and Modernize," *Foreign Policy* (March–April 2015): 52–59.

7 Colin N. Waters et al., "The Anthropocene Is Functionally and Stratigraphically Distinct from the Holocene," *Science* 351, no. 6269 (January 8, 2016); J. Blunden et al., eds., "2015: State of the Climate in 2014," *Bulletin of the American Meteorological Society* 96, no. 7 (July 2015); Elizabeth Kolbert, *The Sixth Extinction* (New York: Henry Holt, 2014); G. Caballos

et al., "Accelerated Modern Human-Induced Species Losses: Entering the Sixth Mass Extinction," *Science* (June 19, 2015); Mark Urban, "Accelerating Extinction Risk from Climate Change," *Science* (May 1, 2015); Johan Rockström et al., "Safe Operating Space for Humanity," *Nature* (September 24, 2009); Rockström and Klum, *Big World, Small Planet*; and Douglas J. McCauley et al., "Marine Defaunation," *Science* (January 16, 2015); Jeremy B. C. Jackson, "Ecological Extinction and Evolution in the Brave New Ocean," *Proceedings of the National Academy of Sciences* (August 12, 2008).

8 AAAS Climate Science Panel, *What We Know: The Reality, Risks and Response to Climate Change* (Washington, DC: American Association for the Advancement of Science, 2013); The Royal Society and National Academy of Sciences, *Climate Change Evidence and Causes* (London: Royal Society, and Washington, DC: NAS, 2014). John Berger, *Climate Peril* (Berkeley, CA: Northbrae Books, 2014) is a very accessible and thorough overview of the science; David Ray Griffin, *Unprecedented: Can Civilization Survive the CO_2 Crisis?* (Atlanta: Clarity Press, 2015), is particularly good on issues of denial and policy changes.

9 Other examples include Lewis Mumford, Norbert Weiner, Jacques Ellul, David Ehrenfeld, Nicholas Carr, Jaron Lanier, Yevgeny Morozov, Lori Andrews, and Frank Pasquale.

10 Wiener is quoted in John Markoff, "In 1949, He Imagined an Age of Robots," *New York Times*, May 21, 2013, p. D8 (emphasis added); Daniel Crevier, *AI: The Tumultuous History of the Search for Artificial Intelligence* (New York: Basic Books, 1993), p. 341.

11 Bill Joy, "Why the Future Doesn't Need Us," *Wired* (April 2000). Physicist Stephen Hawking similarly warns, "Once humans develop artificial intelligence, it would take off on its own and redesign itself . . . the development of full artificial intelligence could spell the end of the human race." Elon Musk likewise warns: "I think we should be very careful about artificial intelligence. If I had to guess at what our biggest existential threat is, it's probably that . . . we are summoning the demon." Quoted in *Time*, December 29, 2014–January 5, 2015. Musk has been quoted as saying that AI is our "biggest existential threat" but is also a major investor in an AI company, OpenAI. See also John Markoff, "Artificial-Intelligence Research Center Is Started by Silicon Valley Investors," *New York Times*, December 11, 2015; Sue Halpern, "How Robots and Algorithms Are Taking Over," *New York Review of Books*, April 2, 2015; Colin Allen, "The Future of Moral Machines," *New York*

Times, December 15, 2011; John Brockman, ed., *What to Think about Machines That Think* (New York: Harper Perennial, 2015).

12 Nick Bostrom, *Superintelligence* (Oxford, Eng.: Oxford University Press, 2014), pp. 105, 259; James Barrat, *Our Final Invention* (New York: Thomas Dunne Books, 2013), pp. 12, 266; George Zarkadakis, *In Our Own Image* (New York: Pegasus Books, 2015), pp. 269, 302, 314, 317.

13 And the pace is accelerating. See, for example, John Markoff, "A Learning Advance in Artificial Intelligence Rivals Human Abilities," *New York Times*, December 10, 2015; Brenen Lake et al., "Human-Level Concept Learning through Probabilistic Program Induction," *Science* 350, no. 6266 (December 11, 2015): 1332–1338; James Martin, *The Meaning of the 21st Century* (New York: Riverhead Books, 2006), p. 223.

14 "The Law of Accelerating Returns predicts a complete merger of the species with the technology it originally created," writes Ray Kurzweil in his book *The Age of Spiritual Machines* (New York: Penguin books, 1999), p. 256; see also Kurzweil, *The Singularity Is Near* (New York: Penguin, 2005); and James Lovelock, *A Rough Ride to the Future* (London: Allen Lane, 2014), p. 167.

15 Zarkadakis, *In Our Own Image*, p. 269.

16 In other words, if we create a system based on threats of mutually assured annihilation, sooner or later, by accident or malice, it will become manifest. Or program an economy for indiscriminate growth at all costs and sooner or later the bill will arrive—as ecological and human ruin. Or idolize machines, and sooner or later, one way or another, they will come to rule. It's in the fine print of all Faustian bargains. There were branch points in the evolution of science, often ignored. The architects of modern science—Galileo, Francis Bacon, and René Descartes— aimed to make nature legible to human intelligence by reducing physical phenomena to their component parts. Implicit in the philosophy of the scientific revolution was a belief that human mastery over nature is a good thing and that its risks and costs would be manageable at an affordable price. The results of growing scientific knowledge included the considerable advantages of improved health, comfort, mobility, greater wealth, and less drudgery. But the revolution in science also created the means by which war became total, laying the ground for the killing fields of the twentieth century in which 200 million people were slaughtered. It created the means by which pollution became a global scourge and climate destabilization became a stark reality. More difficult to see and so more insidious, was its effect on how we thought and what we thought

about. Scientific reductionism tended to limit our thinking to fragments, making it difficult to fathom the larger systems in which we live and to foresee the long-term consequences of our actions. Note E. F. Schumacher's distinction between "convergent" and "divergent" problems in Schumacher, *A Guide for the Perplexed* (New York: Harper and Row, 1977), pp. 120–136; as well as David W. Orr, "Catastrophe and Social Order," *Human Ecology* (March 1979).

17 In particular we need to create reasonable and enforceable policy and laws that improve the efficiency with which we use energy, water, and materials, and that advance renewable energy. For example, Constitutional scholars Lawrence Tribe and Joshua Matz write: "The Supreme Court is part of the world it helps create. No major case is truly over when it's over. The winners and losers may go home, but the reverberations from the Court's decision continue, generating new controversies that will return to a Court that has itself been changed by the altered legal and cultural landscape." Tribe and Matz, *Uncertain Justice* (New York: Henry Holt, 2014), p. 315. When asked in 1957 what he thought about the French Revolution of 1792, Chinese premier Chou En Lai famously replied, "It is too soon to tell."

18 Criticisms of democracy began with that of Plato after the execution of Socrates. More recent critics include that engineer of public consent, Edward Bernays, the skeptical economist, Joseph Schumpeter, that all-purpose curmudgeon, H. L. Mencken, and finally that indefatigable molester of civic intelligence, Rupert Murdoch. For various reasons, critics of democracy were mostly blind to similar traits exhibited by elites and by the best and brightest of their times. See also Bryan Caplan, *The Myth of the Rational Voter* (Princeton, NJ: Princeton University Press, 2007); Beth Simone Noveck, *Smart Citizens, Smarter State* (Cambridge, MA: Harvard University Press, 2015), pp. 75–99.

Chapter 2. The Challenges of Sustainability

Epigraphs. Nicholas Georgescu-Roegen, *The Entropy Law and the Economic Process* (Cambridge, MA: Harvard University Press, 1971), p. 304; Andreas Malm, *Fossil Capital* (London: Verso, 2016), p. 393.

1 International Union for the Conservation of Nature, *World Conservation Strategy* (Gland, Switz.: IUCN, 1980); Lester Brown, *Building a Sustainable Society* (New York: Norton, 1981); Gro Harlem Brundtland et al., *Our Common Future* (New York: Oxford University Press, 1987);

Jeremy Caradonna, *Sustainability: A History* (New York: Oxford University Press, 1914), p. 19; Leslie Paul Thiele, *Sustainability* (Cambridge, Eng.: Polity Press, 2013), p. 5; Brendon Larson, *Metaphors for Environmental Sustainability* (New Haven: Yale University Press, 2011), p. 99; Peter Jacques, *Sustainability: The Basics* (London: Routledge, 2015).

2 Michael Lewis and Pat Conaty, *The Resilience Imperative* (Gabriola Island, BC: New Society Publishers, 2012). Kent Portney, *Sustainability* (Cambridge, MA: MIT Press, 2015) is a very useful and succinct guide to the subject, particularly on issues relating to governance.

3 Diane Ackerman, *The Human Age* (New York: Norton, 2014), pp. 308–309; Jane Jacobs, *Dark Age Ahead* (New York: Random House, 2004), pp. 4, 24; Clive Hamilton, *Requiem for a Species* (London: Earthscan, 2010), pp. 211, 218, 223.

4 Paul Kingsnorth, "Dark Ecology," *Orion* (January–February 2013): 18–29; Daniel Smith, "It's the End of the World as We Know It . . . and He Feels Fine," *New York Times Magazine*, April 17, 2014. Paul Gilding, in his book *The Great Disruption* (New York: Bloomsbury, 2011), makes the case that we will prevail over climate change because we must.

5 Elizabeth Kolbert, *The Sixth Extinction* (New York: Henry Holt, 2014). For example, the Drawdown Project organized by Paul Hawken; see Tim Flannery, *Atmosphere of Hope* (New York: Atlantic Monthly Press, 2015), p. 175. Timothy Snyder, in his *Black Earth: The Holocaust as History and Warning* (New York: Tim Buggan Books, 2015), argues that climate change could usher in another holocaust, only worse.

6 Estimates of carbon uptake and retention vary by an order or two of magnitude—email from Wes Jackson at the Land Institute, November 23, 2015. The same, I think, will be true of calculations about the energy return on the investment, which in some proposals will require making and transporting billions of soil amendments like biochar every year. And there is the perennial issue of hyping technologies that do not and cannot meet tests of net energy, duration, scale, time, cost, and practical implementation but never fail to arouse hopes that migrate from one proposed breakthrough to another. The climate solutions business, however, is in business to sell solutions.

7 Alberto Manguel, *The City of Words* (Toronto: Anansi Press, 2007), pp. 136, 141.

8 Edward Bernays, *Propaganda* (1928; Brooklyn: Ig Publishing, 1955), p. 37; also Edward Bernays, *Crystallizing Public Opinion* (1923; Brooklyn: Ig Publishing, 1951); and Stuart Ewen, *Captains of Consciousness* (New York:

McGraw-Hill, 1976). Jill Lapore, "The Lie Factory," *New Yorker*, September 24, 2012, pp. 50–59, describes the first political consulting firm of Whitaker and Baxter. It may be, as Lapore writes, that "advertising began as a form of political consulting," not the other way around. At the end of her ninety-five years, Leone Baxter is quoted as saying that political consulting, a field which she helped to create, "could be a very, very destructive thing"; see Clive Hamilton, *Requiem for a Species* (London: Earthscan, 2010), pp. 87–88.

Chapter 3. Resilience

Epigraph. Václav Havel, *Disturbing the Peace* (New York: Vintage, 1991), p. 16.

1 Nicholas Taleb, *The Black Swan*, 2nd ed. (New York: Random House, 2010); and Taleb, *Antifragile* (New York: Random House, 2012). In the latter, Taleb writes: "The modern world may be increasing in technological knowledge, but, paradoxically, it is making things a lot more unpredictable. Now for reasons that have to do with the increase of the artificial, the move away from ancestral and natural models, and the loss in robustness owing to complication in the design of everything, the role of Black Swans is increasing" (p. 7).

2 Lance H. Gunderson and C. S. Holling, eds., *Panarchy* (Washington, DC: Island Press, 2002); C. S. Holling, "Understanding the Complexity of Economic, Ecological, and Social Systems," *Ecosystems* 4 (2001): 390–405; also Brian Walker and David Salt, *Resilience Thinking* (Washington, DC: Island Press, 2006).

3 Amory Lovins and Hunter Lovins, *Brittle Power* (Andover, MA; Brick House, 1984), particularly chapter 13; Edward Tenner, *Why Things Bite Back: Technology and the Revenge of Unintended Consequences* (New York: Knopf, 1996); Charles Perrow, *Normal Accidents* (Princeton, NJ: Princeton University Press, 1999); Charles Perrow, *The Next Catastrophe* (Princeton, NJ: Princeton University Press, 2007); Jacques Ellul, *The Technological Society* (New York: Vintage Books, 1964); and Jacques Ellul, *The Technological System* (New York: Continuum, 1980).

4 World Economic Forum, *Global Risks, 2016* (Geneva, Switz.: World Economic Forum, 2016); Martin Rees, "Denial of Catastrophic Risks," *Science* 339 (March 8, 2013), p. 1123. At a lower level, much of our infrastructure, like the electrical grid, is vulnerable to acts of God, incompetence, and malice. For instance read Ted Koppel, *Lights Out: A Cyberattack, a Nation Unprepared, Surviving the Aftermath* (New York: Crown, 2015); National

Research Council, *Terrorism and the Electric Power Delivery System* (Washington, DC: NRC, 2012); U.S. Department of Energy, *U.S. Energy Sector Vulnerabilities to Climate Change and Extreme Weather* (Washington, DC: USDOE, 2013); National Academy of Sciences, *Disaster Resilience* (Washington, DC: The National Academies Press, 2012).

5 Ronald Wright, *A Short History of Progress* (New York: Carroll & Graf, 2005), pp. 7, 8, 108; Robert Costanza, "Social Traps and Environmental Policy," *Bioscience* (June 1987): 407–412; and Taleb, *Antifragile.*

6 Bill Joy, "Why the Future Doesn't Need Us," *Wired* (April 2000), note 35. Cost-benefit analysis, improperly applied, is like using a sledge hammer to do dental work. It economizes on subtlety and nuance. See Frank Ackerman and Lisa Heinzerling, *Priceless* (New York: New Press, 2004); and Kerry Whiteside, *Precautionary Politics* (Cambridge, MA: MIT Press, 2006).

7 Jules Pretty, "The Consumption of a Finite Planet," *Environment & Resource Economics* (May 2013); Richard Wilkinson and Kate Pickett, *The Spirit Level* (London: Penguin Books, 2010).

8 Peter Burnell, *Climate Change and Democratization* (Berlin: Heinrick Böll Stiftung, 2009); also Joshua Kurlantzick, *Democracy in Retreat* (New Haven: Yale University Press, 2013).

9 John A. Livingston, *Rogue Primate* (Toronto: Key Porter Books, 1994).

10 Donella Meadows, *Thinking in Systems* (White River Junction, VT: Chelsea Green, 2008), p. 76; Peter Fox-Penner, *Smart Power* (Washington, DC: Island Press, 2010).

11 Eric Klinenberg, "Adaptation," *New Yorker* (January 7, 2013): 35.

12 Mark Mykleby and Wayne Porter [authors called themselves "Mr. Y," which is reminiscent of George Kennan's "X" article in 1947], *A National Strategic Narrative* (Washington, DC: Woodrow Wilson Center, 2011); Mark Mykleby, Patrick Doherty, and Joel Makower, *The New Grand Strategy: Prosperity, Security, and Sustainability in the 21st Century* (New York: St. Martin's, 2016); Barry Lynn, *End of the Line* (New York: Doubleday, 2005), p. 234; Barry Lynn, "Built to Break," *Challenge* (March–April 2012): 87–107; also Andrew Winston, "Resilience in a Hotter World," *Harvard Business Review* (April 2014): 56–92; Perrow, in *Next Catastrophe*, p. 295, writes that "smaller organizations have a smaller potential for harm" and so proposes reducing the size of organizations. See also Patrick Doherty, "A New U.S. Grand Strategy," *Foreign Policy*, available online at http://foreignpolicy.com/2013/01/09/a-new-u-s-grand-strategy (accessed February 23, 2016).

13 Andrew Zolli, "Learning to Bounce Back," *New York Times*, November 2, 2012; Nicholas Taleb, *Antifragile*, p. 7.

14 Taleb, *Antifragile*, p. 285.

15 Al Gore, *The Future* (New York: Random House, 2013), p. xxv.

16 Nicholas Berggruen, *Intelligent Governance for the 21st Century* (Cambridge, Eng.: Polity Press, 2012), pp. 13, 34, 182; Leon S. Fuerth, "Anticipatory Governance: Practical Upgrades," October 2012, p. 1, unpublished ms.

17 Neil Postman, *Building a Bridge to the Eighteenth Century* (New York: Knopf, 2000).

Chapter 4. The Problem of Denial

Epigraph. Nicole Krauss, *The History of Love* (New York: W.W. Norton, 2005), p. 8.

1 Svante Arrhenius, "On the Influence of Carbonic Acid in the Air upon the Temperature of the Ground," *The London, Edinburgh, and Dublin Philosophical Magazine and Journal of Science* 41, no. 251 (April 1896); Spencer Weart, *The Discovery of Global Warming* (Cambridge, MA: Harvard University Press, 2003); George Woodwell's *A World to Live In* (Cambridge, MA: MIT, 2016) is a remarkably clear and authoritative review of the current science and policy challenges. The quote is from a private email sent to me by George Woodwell, January 6, 2016. World Meteorological Organization, "Greenhouse Gas Bulletin," November 9, 2015; World Meteorological Organization, press release, November 26, 2015. Between 2008 and 2014, the rate of solar deployment was about 50 percent a year. See Dickon Pinner and Matt Rodgers, "Solar Power Comes of Age," *Foreign Affairs* (March–April 2015): 111–118; *National Climate Assessment, 2014* (Washington, DC: National Academies Press, 2014). It may be possible, as Christiana Figueres, executive secretary of the UN Framework Convention on Climate Change, stated prior to the Paris climate meeting, to hold global temperatures below an increase of two degress Celsius, but it is not likely. The best projections based on national commitments presently are in the 2.7 to 4 degrees Celsius range, and that presumes no nasty surprises from carbon-cycle feedbacks, which would itself be surprising. Johan Rockström and Mattias Klum, *Big World, Small Planet* (New Haven: Yale University Press, 2015), p. 88. CO_2 levels at 402 ppm in March 2016 are higher than in probably several million years. But the effects are amplified

by other heat-trapping gases such as CH_4 that measured in roughly CO_2 equivalent numbers push the level to ~450–470 ppm CO_{2e}. See Kevin Anderson and Alice Bows, "Beyond 'Dangerous' Climate Change," *Philosophical Transactions of the Royal Society A* 369 (2011): 20–44; Anthony D. Baranosky et al., "Approaching a State Shift in Earth's Biosphere," *Nature* 486 (June 7, 2012): 52–58. Martin Rees of Cambridge University gives civilization only a 50–50 chance of surviving to the year 2100; see Rees, *Our Final Hour* (New York: Basic Books, 2003). He later reported, in *From Here to Infinity* (London: Profile Books, 2011), that his scientist colleagues thought him too optimistic (pp. 72–73).

2 David Archer, *The Long Thaw* (Princeton, NJ: Princeton University Press, 2009); Susan Solomon et al., "Irreversible Climate Change Due to Carbon Dioxide Emissions," *Proceedings of the National Academy of Sciences* 106, no. 6 (February 10, 2009): 1704–1709; or in accessible English, Curt Stager, "Tales of a Warmer Planet," *New York Times*, November 29, 2015; and Curt Stager, *Deep Future* (New York: Thomas Dunne Books, 2011), pp. 29–48.

3 Intergovernmental Panel on Climate Change, *Summary for Policy Makers: Physical Science Basis* (Geneva: IPCC, 2013); David Archer, *The Long Thaw* (Princeton, NJ: Princeton University Press, 2009), p. 1; also James Hansen et al., "Assessing Dangerous Climate Change," www.plosone .org, vol. 8, no. 12 (December 2013), where he writes that "most fossil fuel carbon will remain in the climate system more than 100,000 years" (p. 21). See also Susan Solomon et al., "Irreversible Climate Change Due to Carbon Dioxide Emissions," *Proceedings of the National Academy of Sciences* 106, no. 6 (February 10, 2009), pp. 1704–1709; Global Challenges Foundation, *12 Risks That Threaten Human Civilization* (Oxford, Eng.: Global Challenges Foundation, 2015); and Elizabeth Kolbert, "Unsafe Climates," *New Yorker* (December 7, 2015), p. 24. Consider, for example, Marshall Burke, Solomon Hsiang, and Edward Miguel, *Climate and Conflict*, NBER working paper 20598 (Cambridge, MA: National Bureau of Economic Research, 2014); John Steinbruner et al., *Climate and Social Stress* (Washington: National Research Council, 2013). The full issue of *Science* published August 2, 2013, is a good review of the climate effects on terrestrial ecosystems, marine ecology, species extinctions, food security, diseases, and consequences of sea-ice decline.

4 Much of this we cannot foresee. Consider James W. C. White et al., *Abrupt Impacts of Climate Change* (Washington, DC: National Research Council, 2013); and Richard Monastersky, "Life—A Status Report,"

Nature 516 (December 11, 2014): 159–161. We can also expect massive changes in the oceans. Lisa-Ann Gershwin, author of *Stung!* (Chicago: University of Chicago Press, 2013), concludes that we passed the tipping point for the oceans long ago; see also Jeremy B. C. Jackson, "Ecological Extinction and Evolution in the Brave New Ocean," *Proceedings of the National Academy of Sciences* (August 12, 2008); Douglas McCauley et al., "Marine Defaunation: Animal Loss in the Global Ocean," *Science* 347, no. 6219 (January 16, 2015). Ocean acidification, a related phenomenon, is occurring ten times faster than it did around the last mass marine extinction. See Andy Ridgwell and Daniela Schmidt, "Past Constraints on the Vulnerability of Marine Calcifiers to Massive Carbon Dioxide Release," *Nature Geoscience* 3 (2010): 196–200. Kerry A. Emanuel, "Downscaling CMIP5 Climate Models Shows Increased Tropical Cyclone Activity over the 21st Century," *Proceedings of the National Academy of Sciences* (June 2013); Benjamin Cook et al., "Unprecedented 21st Century Drought Risk in the American Southwest and Central Plains," *Science Advance* (February 12, 2015); Colin P. Kelley et al., "Climate Change in the Fertile Crescent and Implications of the Recent Syrian Drought," *Proceedings of the National Academy of Sciences* (January 30, 2015); Steven Smith et al., "Near-Term, Acceleration in the Rate of Temperature Change," *Nature Climate Change* (April 2015); Kevin E. Trenberth, "Changes in Precipitation with Climate Change," *Climate Research* (January 3, 2011).

5 Susan Solomon et al., "Irreversible Climate Change Due to Carbon Dioxide Emissions," *Proceedings of the National Academy of Sciences* 106, no. 6 (February 10, 2009): 1704–1709. Methane hydrates are suspected as the culprit in the Permian extinction some 250 million years ago that destroyed 90 percent of life on Earth. See Michael Benton, *When Life Nearly Died* (London: Thames & Hudson, 2003). The beast is waking. See Jorgen Hollesen et al., "Permafrost Thawing in Organic Arctic Soils Accelerated by Ground Heat Production," *Nature Climate Change* (April 6, 2015); M. O. Clarkson et al., "Ocean Acidification and the Permo-Triassic Mass Extinction," *Science* 348, no. 6231 (April 10, 2015), pp. 229–232.

6 Broecker's widely quoted comment is from a talk he gave at the University of New Mexico in 1991. See also Hans Joachim Schellnhuber, ed., *Tipping Elements in Earth Systems*, special issue of *Proceedings of the National Academy of Sciences* 106, no. 49 (December 2009).

7 The term originated in W. J. Rittel and Melvin Webber, "Dilemmas in a General Theory of Planning," *Policy Sciences* (1973). For perspective, see

Peter J. Balint et al., *Wicked Environmental Problems* (Washington, DC: Island Press, 2011); U.S. Department of Defense, *2014 Climate Change Adaptation Roadmap* (Washington, DC: USDOD, 2014). Climate change will likely exacerbate civil conflicts. See Solomon M. Hsiang et al., "Civil Conflicts Are Associated with the Global Climate," *Nature* 476, no. 25 (August 25, 2011), pp. 438–441; Jonathan Spaner and Hillary LeBail, "The Next Security Frontier," *Proceedings of the U.S. Naval Institute* 139 (October 2013): 30–35; Kurt Campbell et al., *The Age of Consequences: The Foreign Policy and National Security Implication of Global Climate Change* (Washington, DC: Center for Strategic and International Studies, 2007); Sherry Goodman, *National Security and the Threat of Climate Change* (Washington: CNA Corporation, 2007); Testimony of Dr. Thomas Fingar, U.S. Congress, Select Committee on Energy Independence and Global Warming, "National Intelligence Assessment on the National Security Implications of Global Climate Change to 2030," *Congressional Record*, June 25, 2008; U.S. Department of Defense, *Quadrennial Defense Review Report* (Washington, DC: USDOD, 2006, 2010, 2014); Global Challenges Foundation, *12 Risks That Threaten Human Civilization*; Johan Rockström et al., "Planetary Boundaries: Exploring the Safe Operating Space for Humanity," *Ecology and Society* 14, no. 2 (2009); Johan Rockström et al., "A Safe Operating Space for Humanity," *Nature* 461, no. 24 (September 2009): 472–475.

8 I'm not concerned here with climate "deniers" or "skeptics." Most will be greatly instructed by coming climate-change-driven droughts, heat waves, flooding, blights, and otherwise inexplicable events, and will come to accept reality. The rest will perhaps find solace as branch members of the Flat Earth Society, which still keeps the lights on for them at their office in London. In terms of Earth warming, again, we are presently slightly over one degree Celsius of increased heat and are "committed" to another approximately half a degree of warming, which is to say that we may have passed our margin of safety. George Marshall, *Don't Even Think about It* (New York: Bloomsbury, 2014) is a good summation of the various issues involved, not the least of which is "terror management" regarding our own mortality; more to the point, however, is Robert Brulle's study of the role of conservative foundations in bankrolling the climate denial movement. See Brulle, "Institutionalizing Delay," *Climate Change* (December 21, 2013).

9 Naomi Oreskes and Erik Conway, in their *Merchants of Doubt* (New York: Bloomsbury, 2010), show that many of the same persons who

opposed action on ozone depletion, acid rain, and cigarette smoking are involved in climate denial and funded by some of the same corporate sponsors and advocacy think tanks. Not the least of the problem is the $5.3 trillion in annual subsidies for the fossil-fuel industries world-wide, $699 billion of which is from the U.S. government. See David Coady et al., "How Large Are Global Energy Subsidies?" International Monetary Fund working paper WP/15/105 (May 2015). Data from the Yale University Project on Climate Communication, however, show that 63 percent of the public believes climate change is happening, but only 48 percent believe that human actions are the cause.

10 Elizabeth Kolbert, *Field Notes from a Catastrophe* (New York: Blooms-bury, 2006), p. 187. In a nutshell, carbon can be removed from air, but not in ways that are carbon neutral, affordable, and deployable at a scale and rate that would significantly affect the problem. Merely to break even, any such device would have to remove about nine billion tons of carbon each year, which would somehow have to be sequestered perma-nently, otherwise there would be no point in making the effort. There is disagreement about whether renewable energy and advanced effi-ciency could support our present manner of living. Kentucky writer Wendell Berry, for one, says that "we will have to live more poorly than we do." At the other end of the spectrum, Amory Lovins believes that if properly powered and rendered efficient we could continue much as we have.

11 Marshall, *Don't Even Think about It.*

12 Richard Hofstadter, *Anti-Intellectualism in American Life* (New York: Vintage, 1963); Al Gore, *The Assault on Reason* (New York: Penguin Press, 2007); and Susan Jacoby, *The American Age of Unreason* (New York: Pantheon, 2008).

13 T. S. Eliot, *The Complete Poems and Plays* (New York: Harcourt, Brace, and World, 1971), p. 118; Barbara Ehrenreich, *Bright-Sided* (New York: Metropolitan Books, 2009), p. 11.

14 Fyodor Dostoyevsky, *The Brothers Karamazov* (New York: The Modern Library, 1950), p. 300.

15 Brulle, "Institutionalizing Delay." Justin Farrell, "Corporate Funding and Ideological Polarization about Climate Change," *Proceedings of the National Academy of Sciences* (November 2015) documents the perverse effects of oil money in funding the polarization of public attitudes on climate; see also Thomas Dietz et al., "Political Influences on Green-house Gas Emissions from U.S. States," *Proceedings of the National Academy*

of Sciences (June 2014); and Lorien Jasny et al., "An Empirical Examination of Echo Chambers in U.S. Climate Policy Networks," *Nature Climate Change* 5 (August 2015). Kari Marie Norgaard, *Living in Denial* (Cambridge, MA: MIT Press, 2011) is a revealing case study that concludes with the question, "How do we break through denial into awareness?" (p. 227). The possible answers are troubling.

16 Rebecca Solnit describes it this way in *A Paradise Built in Hell* (New York: Viking, 2009).

17 A report by The American Physical Society (2011) regarded the challenge as "formidable" and not useful in the near term. The challenges include (1) developing reliable technology to remove the annual flow of carbon to the atmosphere and more in ways that are carbon neutral, (2) permanently storing a very large volume of carbon extracted, either underground without causing earthquakes or in materials, (3) taking action at a scale commensurate with the problem, and at a price we are willing to pay, and (4) doing these things before all Hell breaks loose. See Robert Socolow et al., *Direct Air Capture of CO_2 with Chemicals* (Washington, DC: American Physical Society, 2011); also John Collins Rudolf, "Physicist Group's Study Raises Doubts on Capturing Carbon Dioxide from Air," *New York Times*, May 10, 2011. David Biello is equally skeptical about capturing carbon from coal plants; see Biello, "The Carbon Capture Fallacy," *Scientific American* (January 2016): 59–65; Clive Hamilton, *Earthmasters* (New Haven: Yale University Press, 2013); Oliver Morton, *The Planet Remade* (Princeton, NJ: Princeton University Press, 2015); National Academy of Sciences, *Climate Intervention* (Washington DC: NAS, 2015). Both are highly skeptical of the "ready, fire, aim" strategy of geoengineering.

18 The phrase is that of Lisa-Ann Gershwin (see her *Stung!*, p. 338). Barbara Ehrenreich calls "defensive pessimism" the trait that keeps us alert to dangers; see Ehrenreich, *Bright-Sided*, p. 200.

Chapter 5. Economy

Epigraphs. William Shakespeare, *King Lear*, 4.1.3; Henry David Thoreau, *Walden* (Princeton, NJ: Princeton University Press, 2004), chapter 1.

1 Ray Kurzweil, *The Singularity Is Near* (New York: Penguin, 2005); Benjamin R. Barber, *Consumed* (New York: Norton, 2007), pp. 3–37; see also Daniel Bell, *The Cultural Contradictions of Capitalism* (New York: Basic Books, 1976).

NOTES TO PAGES 48-49

2 A. C. Pigou, *The Economics of Welfare* (London: Macmillan, 1920); John Kenneth Galbraith, *The Essential Galbraith* (Boston: Mariner Books, 2001); Kenneth Boulding, "Economics of the Coming Spaceship Earth," in H. Jarrett, ed., *Environmental Quality in a Growing Economy* (Baltimore: Johns Hopkins University Press, 1966); Robert Heilbroner, *An Inquiry into the Human Prospect* (1974; New York: Norton, 1980); Nicholas Georgescu-Roegen, *The Entropy Law and the Economic Process* (Cambridge, MA: Harvard University Press, 1974); Herman Daly, *Ecological Economics and Sustainable Development* (Cheltenham, Eng.: Edward Elgar, 2007); Herman Daly, *From Uneconomic Growth to a Steady-State Economy* (Cheltenham, Eng.: Edward Elgar, 2014); Herman Daly and John B. Cobb, *For the Common Good* (1989; Boston: Beacon Press, 1994). In Robert Skidelsky and Edward Skidelsky's words, "Economics is not just any academic discipline. It is the theology of our age, the language that all interests, high and low, must speak if they are to win a respectful hearing in the courts of power." They attribute this to the "failure of other disciplines to impress their stamp on political debate" (New York: Other Press, 2012), p. 92. Naomi Klein writes that the belief "that we are nothing but selfish, greedy, self-gratification machines . . . is neoliberalism's single most damaging legacy." Klein, *This Changes Everything* (New York: Simon & Schuster, 2014), p. 62; Gary Becker, *The Economic Approach to Human Behavior* (Chicago: University of Chicago Press, 1976), p. 8.

3 David Harvey, *A Brief History of Neoliberalism* (Oxford, Eng.: Oxford University Press, 2005), pp. 2–3, 159; Thatcher quoted in Naomi Klein, *This Changes Everything*, p. 60.

4 The sensible commentators include Hunter Lovins and Boyd Cohen, *Climate Capitalism* (New York: Hill and Wang, 2011); Marjorie Kelly, *Owning Our Future* (San Francisco: Berrett-Koehler, 2012); and Klein, *This Changes Everything*. Better methods of accounting would certainly help, as Jane Gleeson-White writes in *Six Capitals* (Sydney: Allen & Unwin, 2014), but there is, she writes, "a logical inconsistency at the heart of the six capitals model which will prevent it alone from saving the planet: it seeks to account for non-financial value but can see it only in terms of financial value. This is because the entity it seeks to govern, the corporation as we know it, is legally bound to make decisions in favour of financial capital" (p. 282). Robert Reich finds little evidence that corporations will be "socially responsible, at least not to any significant extent." Further, what often pass for social responsible actions, according to Reich,

are nothing more than gussied up efforts to reduce costs. See Robert Reich, *Supercapitalism* (New York: Knopf, 2007), pp. 170–171; also Robert Skidelsky and Edward Skidelsky, *How Much Is Enough?* (New York: Other Press, 2012), p. 101.

5 David Colander and Roland Kupers, *Complexity and the Art of Public Policy* (Princeton, NJ: Princeton University Press, 2014), p. 276.

6 Quoted in Ted Nace, *Gangs of America* (San Francisco: Barrett-Koehler, 2003), p. 15; Morton J. Horwitz, *The Transformation of American Law, 1780–1860* (Cambridge, MA: Harvard University Press, 1977), pp. 253–254.

7 Nace, *Gangs of America*, pp. 102–117; also Marjorie Kelly, *The Divine Right of Capital* (San Francisco: Berrett-Koehler, 2001), quotation on p. 163. The idea of corporate personhood, according to attorney Tom Linzey, has antecedents in English Common Law (private conversation, November 14, 2015).

8 *New York Times*, October 23, 2008. p. 1; Alan Greenspan, "Never Saw It Coming," *Foreign Affairs* 92, no. 6 (November–December 2013): 88–96. Greenspan claims that no one saw the 2008 crisis coming and otherwise blames it on "animal spirits," that is, the irrationality of everyone else; Paul Krugman is quoted in David Orrell, *Economyths: Ten Ways Economics Gets It Wrong* (Ontario: John Wiley & Sons, 2010), p. 106; Tony Judt, *Ill Fares the Land* (New York: Penguin, 2010), p. 2.

9 That includes bridges, roads, water systems, schools that educate our children, public transportation, and might have also included clean, efficient, high-speed rail systems found to be efficacious for many of the countries that we whupped up on (a useful phrase from Muhammed Ali) in bygone wars; Judt, *Ill Fares the Land*, p. 224.

10 The underwhelming pedagogical accomplishments of economists has been noted by their students, who have organized under the banner "the International Student Initiative for Pluralist Economics" to assist their professors in forming a discipline better suited to the classroom and to the exigencies of the twenty-first century. See Phillip Inman, "Economics Students Call for a Shakeup of the Way Their Subject Is Taught," *The Guardian*, May 4, 2014; Ivan Illich, *Medical Nemesis* (New York: Bantam Books, 1976).

11 Vaclav Smil, *Power Density* (Cambridge, MA: MIT Press, 2015); and Benjamin Sovacol, *The Dirty Energy Dilemma* (Westport, CT: Praeger, 2008).

12 Bill McKibben, "Climate Change's Terrifying New Math," *Rolling Stone* (August 2, 2012).

13 Beef is the worst for lots of reasons. Read Denis and Gail Boyer Hayes, *Cowed* (New York: Norton, 2015); Benjamin Cook et al., "Unprecedented 21st Century Drought Risk in the American Southwest and Central Plains," *Science* (February 12, 2015); Colin P. Kelly et al., "Climate Change in the Fertile Crescent and Implications of the Recent Syrian Drought," *Proceedings of the National Academy of Science* 112, no. 11 (2015).

14 Colin Price, *Time, Discounting and Value* (Oxford, Eng.: Blackwell, 1993). This is a persuasive and eloquent case that "discounting the value of future goods, services, resources, events and experiences, by means of a uniform negative exponential function, cannot be justified" (p. 345); Nicholas Stern, *Why Are We Waiting?* (Cambridge, MA: MIT Press, 2015), pp. 152–184; William Nordhaus, *The Climate Casino* (New Haven: Yale University Press, 2013), pp. 182–194. Their positions differ in the sense of their urgency and their devotion to neoclassical economic dogma. My own more jaundiced thoughts on the relation of the profession to biophysical realities can be found in "Pascal's Wager and Economics in a Hotter Time," *Earth in Mind* (1994; Washington, DC: Island Press, 2004).

15 David Ricardo, for example, once described "the original and indestructible powers of the land alongside other gifts of nature which exist in boundless quantity." Quoted in Gilbert Rist, *The Delusions of Economics* (London: ZED Books, 2011), p. 171.

16 John Gowdy, ed., *Limited Wants, Unlimited Means: A Reader on Hunter-Gatherer Economics and the Environment* (Washington, DC: Island Press, 1998); Marshall Sahlins, *Stone Age Economics* (Chicago: Aldine, 1972). Colander and Kupers similarly believe "it is useful for society to keep some standard economists around to remind society of fundamental rules. But it hardly justifies an entire scientific discipline which should aim at a more comprehensive understanding." See Colander and Kupers, *Complexity and the Art of Public Policy*, p. 73.

17 Herman Daly is the most authoritative and accessible economist on the subject. Anthony Baranosky et al., "Approaching a State Shift in Earth's Biosphere," *Nature*, 486 (June 2012): 52–58. Since we are more ignorant than smart and probably always will be, the subject of ignorance should be studied closely by the learned. Other than, say, the book of Ecclesiastes, the best book on the subject is Bill Vitek and Wes Jackson, eds., *The Virtues of Ignorance* (Lexington: University of Kentucky Press, 2008); Robert Sinsheimer, "The Presumptions of Science," *Daedalus* 107, no. 2 (Spring 1978): 23–35.

18 John Stuart Mill, *Principles of Political Economy* (1848; London: Longman, Green, and Co., 1940), pp. 746–751; Donella Meadows et al., *The Limits to Growth* (New York: Universe Books, 1972); Donella Meadows et al., *Limits to Growth: The 30-Year Update* (White River Junction, VT: Chelsea Green, 2004); Graham Turner, "A Comparison of the Limits to Growth with Thirty Years of Reality" (Canberra, Austr.: CSIRO, 2008); see also Richard Heinberg's superb *The End of Growth* (Gabriola Island, BC: New Society, 2011); Kerryn Higgs, *Collision Course: Endless Growth on a Finite Planet* (Cambridge, MA: MIT Press, 2014); Joseph Tainter, *The Collapse of Complex Societies* (Cambridge, Eng.: Cambridge University Press, 1988); Jared Diamond, *Collapse* (New York: Viking, 2005); and Thomas Homer-Dixon, *The Ingenuity Gap* (New York: Knopf, 2000).

19 U. Thara Srinivasan et al., "The Debt of Nations and the Distribution of Ecological Impacts from Human Activities," *Proceedings of the National Academy of Sciences*, 105, no. 5 (February 5, 2008): 1768–1773; Juliet Schor, *Plenitude* (New York: Penguin, 2010), p. 94.

20 Bob Johnson, *Carbon Nation* (Lawrence: University Press of Kansas, 2014), pp. 5, 12.

21 Richard Heinberg, *The Party's Over* (Gabriola Island, BC: New Society, 2003). Richard Heinberg, *Snake Oil* (Santa Rosa, CA: Post-Carbon Institute, 2013), p. 29; Heinberg, *Afterburn: Society beyond Fossil Fuels* (Gabriola Island, BC: New Society, 2015), pp. 23–28; Vaclav Smil, *Power Density* (Cambridge, MA: MIT Press, 2015), pp. 254–255. Smil estimates the EROI necessary to maintain modern civilization at 12 to 14, which he believes is higher than possible anytime soon with solar and wind. The conflicts for "decades to come" may be with people who refuse to fight by the Marquis of Queensbury's rules of fair play. See Michael Klare, *The Race for What's Left* (New York: Metropolitan Books, 2012); and Johnson, *Carbon Nation*, pp. 173–174.

22 Amory Lovins et al., *Reinventing Fire* (Snowmass, CO: Rocky Mountain Institute); Ozzie Zehner, *Green Illusions* (Lincoln: University of Nebraska Press, 2012), p. 169. In either case, "getting off fossil fuels . . . will be one of the most difficult challenges modern civilization has ever faced, and it will require the most sustained, well-managed, globally cooperative effort the human species has ever mounted." Gernot Wagner and Martin Weitzman, *Climate Shock: The Economic Consequences of a Hotter Planet* (Princeton, NJ: Princeton University Press, 2015), p. ix.

In *Our Only World* (Berkeley: Counterpoint, 2015), Wendell Berry writes, "We must understand that fossil fuel energy must be replaced not just by 'clean' energy, but also by *less* energy. The unlimited use of *any* energy would be as destructive as unlimited economic growth" (p. 71). Ted Trainer, *Renewable Energy Cannot Sustain a Consumer Society* (Dordrecht, Neth.: Springer, 2007), p. 7.

23 Clive Hamilton, *Growth Fetish* (London: Pluto Press, 2004), pp. 91, 219. Robert Kennedy, speech at Kansas University, March 18, 1968.

24 Jonathan Rowe, "Our Phony Economy," *Harpers* (June 2008): 17–24, quotation on p. 23. Michael J. Sandel, *What Money Can't Buy* (New York: Farrar, Straus, and Giroux, 2012).

25 Architect Lance Hosey reports that in 1994 "there were half a million different consumer goods for sale in the United States, and now Amazon alone offers 24 million." *The Shape of Green* (Washington, DC: Island Press, 2012), p. 96.

26 The word "pre-polluted" is from the 2008–2009 Annual Report of the President's Cancer Panel, *Reducing Environmental Cancer Risks* (Washington, DC, 2010), p. vii. Francis Bacon, *The New Organon* (Indianapolis: Bobbs-Merrill Co., 1960), p. 25.

27 E. F. Schumacher, *Small Is Beautiful: Economics as if People Mattered* (New York: Harper Torchbooks, 1973), pp. 31, 54; On Gandhian economics, see J. C. Kumarappa, *The Economy of Permanence: A Quest for a Social Order Based on Non Violence* (1946; Delhi: All India Village Industries Association, 2015).

28 Karl Polanyi, *The Great Transformation* (1944; Boston: Beacon Press, 1967), p. 73; Václav Havel, *Living in Truth* (London: Faber and Faber, 1990), p. 62.

29 John Kenneth Galbraith, *Economics, Peace, and Laughter* (New York: Signet Books, 1971), p. 32. Chuck Collins and Josh Hoxie, *Billionaire Bonanza: The Forbes 400 and the Rest of Us* (Washington, DC: Institute for Policy Studies, 2015); "Wealth: Having It All and Wanting More," Oxfam Issue Briefing (January 2015), p. 3.

30 Thomas Piketty, *Capitalism in the Twenty-First Century* (New York: Penguin, 2014), pp. 294, 573; Richard Wilkinson and Kate Pickett, *The Spirit Level: Why Equality Is Better for Everyone* (London: Penguin Books, 2010); also Jill Lepore, "Richer and Poorer," *New Yorker* (March 16, 2015): 26–32; Joseph Stiglitz, *The Great Divide* (New York: Norton, 2015); Anthony Atkinson, *Inequality: What Can Be Done?* (Cambridge, MA: Harvard

University Press, 2015). Jane Jacobs, for example, writes that "Today the Soviet Union and the United States each predicts and anticipates the economic decline of the other. Neither will be disappointed." Jacobs, *Cities and the Wealth of Nations* (New York: Random House, 1984), p. 200.

31 David Harvey, *The Enigma of Capital* (New York: Oxford University Press, 2010), p. 260; see also David Harvey, *Seventeen Contradictions and the End of Capitalism* (New York: Oxford University Press, 2014), in which he writes that "the only hope is that the mass of humanity will see the danger before the rot goes too far" (p. 293). Howard T. Odum and Elisabeth C. Odum, *A Prosperous Way Down* (Boulder: University of Colorado, 2001), p. 4.

32 Gene Logsdon, "Amish Economics," in *Living at Nature's Pace* (White River Junction, VT: Chelsea Green, 2000), pp. 130–140.

33 Ibid., pp. 132–133.

34 In *Amish Grace* the authors tell the moving story of Amish people forgiving the killer and upholding his family following the massacre of ten Amish school girls at Nickel Mines, Pennsylvania, in 2006. The story stands in bold relief against the present tsunami of religious hatreds and violence and counterhatred and violence. See Donald Kraybill et al., *Amish Grace* (New York: John Wiley & Sons, 2007).

35 Richard Louv, *The Last Child in the Woods* (Chapel Hill, NC: Algonquin Books, 2005); Richard Louv, *The Nature Principle* (Chapel Hill, NC: Algonquin Books, 2011).

36 Mark C. Taylor, *Speed Limits* (New Haven: Yale University Press, 2014). See also Peter Toohey, *Boredom* (New Haven: Yale University Press, 2011), particularly where Toohey, in an interesting insight into boredom in our time, argues that it may be helpful in "provid[ing] an early warning signal that certain situations may be dangerous to our well-being" (p. 174).

37 George Sturt, *The Wheelwright's Shop* (1923; Cambridge, Eng.: Cambridge University Press, 1984), pp. 18, 53, 66; Jacquetta Hawkes, *A Land* (New York: Random House, 1951), pp. 143–144. For a description of a craft organmaker along similar lines, see Matthew B. Crawford, *The World beyond Your Head* (New York: Farrar, Straus and Giroux, 2015), pp. 209–246.

38 The history is long and sad. The instances include the Highland clearances in Scotland, as well as the Soviet and Chinese collectivization of peasant farmers. Rees quoted in Leopold Kohr, *The Breakdown of Nations* (1957; New York: Dutton, 1978), p. 221.

39 As the great economist Nicholas Georgescu-Roegen once put it, "Man's nature being what it is, the destiny of the human species is to choose a truly great but brief, not a long and dull, career." See Nicholas Georgescu-Roegen, *The Entropy Law and the Economic Process* (Cambridge, MA: Harvard University Press, 1974), p. 304.

40 Peter Diamandis and Steven Kotler, *Abundance: The Future Is Better than You Think* (New York: The Free Press, 2012), pp. 302–304.

41 James Howard Kunstler, *Too Much Magic* (New York: Atlantic Monthly Press, 2012); Rob Hopkins, *The Transition Towns Handbook* (Totnes, Eng.: Green Books, 2008).

42 Read, for example, James Gustave Speth, *America the Possible: Manifest for a New Economy* (New Haven: Yale University Press, 2012); Gar Alperovitz, *America beyond Capitalism* (Takoma Park, MD: Democracy Collaborative, 2011); Gar Alperovitz and Lew Daly, *Unjust Deserts* (New York: The New Press, 2008); Herman Daly and John B. Cobb, *For the Common Good* (Boston: Beacon Press, 1994); Jeff Gates, *The Ownership Solution* (Cambridge, Eng.: Perseus Books, 1999); Jeff Gates, *Democracy at Risk* (Cambridge, Eng.: Perseus Books, 2000); J. David Korten, *Agenda for a New Economy* (San Francisco: Berrett-Koehler Publishers, 2010); Juliet Schor, *Plenitude* (New York: Penguin Press, 2010); David Schweickart, *After Capitalism* (New York: Rowman & Littlefield, 2002); Michael Shuman, *The Local Economy Solution* (White River Junction, VT: Chelsea Green, 2015); John Ruskin, *"Unto This Last"* (London: Dutton, 1968), pp. 192, 202; Kirkpatrick Sale, *Human Scale* (New York: Coward, McCann, & Geoghegan, 1980); Wolfgang Sachs et al., *Greening the North* (London: ZED Books, 1999); E. F. Schumacher, *Good Work* (New York: Harper Colophon, 1979); Eric Gill, *A Holy Tradition of Working* (Ipswich: Golgonooza Press, 1983). Robert and Edward Skildelsky define advertising as "the organized creation of dissatisfaction" (*How Much Is Enough?*, p. 40). Clive Hamilton proposes to "ban advertising and sponsorship from all public spaces and restricting advertising time on television and radio," as well as to further change the tax laws that make it a deductible business expense; see Clive Hamilton, *Growth Fetish* (London: Pluto Press, 2004), p. 219. Professor Hamilton is exceedingly polite. Dante would have consigned the makers, purveyors, and enablers to one of the lower tiers of Hell. Erich Fromm, *To Have or to Be* (New York: Bantam Books, 1976); David Bollier, *Think Like a Commoner: A Short Introduction to the Life of the Commons* (Gabriola Island, BC: New Society, 2014).

43 The words are from Blaise Pascal, *Pensées*.

Chapter 6. Governance

Epigraphs. James Madison, *Federalist Papers*, No. 51; Brian Barry, *Why Social Justice Matters* (Cambridge, Eng.: Polity Press, 2005), p. 251.

1 Rick Heede, "The Climate Responsibilities of Industrial Carbon Producers" (Cambridge, MA: Union of Concerned Scientists, 2015); for the quotation by Speth, see Letter to the Editor, *New York Times*, May 14, 2014.

2 Edmund Fawcett, *Liberalism: The Life of an Idea* (Princeton, NJ: Princeton University Press, 2013). Precisely what self-described conservative economists wish to conserve is a matter of some dispute. It cannot be the status quo socially or ecologically. One can suppose that it is only the rules of the system by which the already rich are made richer.

3 Jane Mayer, *Dark Money* (New York: Doubleday, 2016); and Zephyr Teachout, *Corruption in America* (Cambridge, MA: Harvard University Press, 2014).

4 Harald Welzer, *Climate Wars: Why People Will Be Killed in the 21st Century* (Cambridge, UK: Polity Press, 2012); Center for Naval Analysis, *National Security and the Threat of Climate Change* (Washington, DC: The Center for Naval Analysis, 2007); The Defense Science Board, *Trends and Implications of Climate Change for National and International Security* (Washington, DC: Office of the Under Secretary of Defense, 2011). As a side note, Henry Kissinger managed to write a lavishly praised book on world order without mentioning climate change, which was rather like writing a review of a play without mentioning the fact that the stage collapsed in the second act. Henry Kissinger, *World Order* (New York: Penguin Books, 2014).

5 George Zarkadakis, *In Our Own Image* (New York: Pegasus Books, 2015), p. 269. Zarkadakis argues that an international ban on AI research is a "very possible scenario" to terminate the multiple threats posed by the advent of artificial intelligence (p. 302).

6 Lisa-Ann Gershwin, *Stung!* (Chicago: University of Chicago Press, 2013).

7 James Hansen et al., "Ice Melt, Sea Level Rise and Superstorms," *Atmospheric Chemistry and Physics Discussions* (2015); Kevin Trenberth et al., "Attribution of Climate Extreme Events," *Nature Climate Change* 5 (2015): 725–730; James Hansen et al., "Perceptions of Climate Change," *Proceedings of the National Academy of Sciences* 109, no. 37 (2012); Mark New et al., "Four Degrees and Beyond," *Philosophical Transactions of the Royal Society* (November 29, 2010).

8   Anthony D. Barnosky et al., "Approaching a State Shift in Earth's Bio-sphere," *Nature* 486 (June 7, 2012).

9   Nicholas Taleb, *The Black Swan* (New York: Random House, 2010); John Casti, *X-Events: The Collapse of Everything* (New York: William Morrow, 2012).

10 Jorgen Randers, *2052: A Global Forecast for the Next Forty Years* (White River Junction, VT: Chelsea Green, 2012), pp. 117–118. Randers writes that by 2052 "there will be more droughts, floods, extreme weather, and insect infestations. The sea levels will be 0.5 meters higher, the Arctic summer ice will be gone and the new weather will bother agriculturalists and vacationers alike . . . and in 2052 the world will be looking with angst toward further change in the second half of the century. Self-reinforcing change will be worry number one—with methane gas emissions from the melting tundra leading to further temperature increase" (pp. 47, 117).

11 Stephen M. Gardiner, *A Perfect Moral Storm: The Ethical Tragedy of Climate Change* (New York: Oxford University Press, 2011). By moral philosopher John Broome's calculations, carbon emissions from those living a normal life in a rich country "wipe out more than six months of a healthy human life. Each year, your annual emissions destroy a few days of healthy life in total. These are serious harms." John Broome, *Climate Matters: Ethics in a Warming World* (New York: Norton, 2012), p. 74.

12 Daniel Kahneman, *Thinking Fast and Slow* (New York: Farrar, Straus and Giroux, 2011); Jonathan Haidt, *The Righteous Mind* (New York: Pantheon, 2012).

13 Mathew Hauer et al., "Millions Projected to Be at Risk from Sea-Level Rise in the Continental United States," *Nature Climate Change* (March 2016).

14 Bill McKibben, "Global Warming's Terrifying New Math," *Rolling Stone* (July 21, 2012). McKibben's estimate is that about $20 trillion in fossil fuel "assets" must be written off if we are to avoid frying the planet. See also www.carbontrackerinitiative.com.

15 Thomas Homer-Dixon cogently describes the many reasons that problems may outrun our ingenuity in *The Ingenuity Gap* (New York: Knopf, 2000); Mark Mazower, *Governing the World* (New York: Penguin, 2012), p. 424.

16 Steven Kelman, "Why Public Ideas Matter," in Robert Reich, ed., *The Power of Public Ideas* (Cambridge, MA: Harvard University Press, 1990), pp. 47–51; Jane Mansbridge, ed., *Beyond Self-Interest* (Chicago: University of Chicago Press, 1990), pp. 3–22.

17 Richard Thaler and Cass Sunstein, *Nudge* (New York: Penguin Books, 2009).

18 Richard K. Matthews, *If Men Were Angels* (Lawrence: University of Kansas Press, 1995), pp. 8, 212.

19 Sanford Levinson, *Our Undemocratic Constitution* (New York: Oxford University Press, 2006); Sanford Levinson, *Framed* (New York: Oxford University Press, 2012); Robert A. Dahl, *How Democratic Is the American Constitution?* (New Haven: Yale University Press, 2002); Steven Hill, *10 Steps to Repair American Democracy* (Sausalito, CA: PoliPoint Press, 2006); Harold Myerson, "Foundering Fathers," *American Prospect* 22, no. 8 (October 2011): 12–17.

20 Charles E. Lindblom, *Politics and Markets* (New York: Basic Books, 1977), p. 356 (emphasis added). But corporations may be misfits in any society in a more fundamental way. If a corporation is a person, as claimed, one could ask what kind of person it is. One possibility is that corporate behavior most often resembles that of a psychopath as portrayed in the 2004 Canadian film *The Corporation*. According to Dr. Robert Hare, an FBI expert on psychopaths, the traits include "a callous unconcern for the feelings of others, incapacity to maintain enduring relationships, reckless disregard for the safety of others, deceitfulness and repeated lying, incapacity to experience guilt, [and] failure to conform to social norms and laws." One might add a willingness to wreck the planet.

21 Richard Lazarus, *The Making of Environmental Law* (Chicago: University of Chicago Press, 2004), pp. 30, 33, 42.

22 Richard Lazarus, "Super Wicked Problems and Climate Change," *Cornell Law Review* 94 (2009): 1153–1234. Reading Jonathan Z. Cannon, *Environment in the Balance* (Cambridge, MA: Harvard University Press, 2015) is cause to wonder what an ecologically and scientifically literate Supreme Court might have done otherwise.

23 Thomas Berry, *Evening Thoughts* (San Francisco: Sierra Club Books, 2006), p. 95.

24 Anthony Giddens, *The Politics of Climate Change* (Cambridge, Eng.: Polity Press, 2009), pp. 91–128.

25 Giddens, *The Politics of Climate Change*, p. 96; Robert Heilbroner, *An Inquiry into the Human Prospect* (1974; New York: Norton, 1980), p. 175. William Ophuls similarly concludes, "Ecological scarcity in particular seems to engender overwhelming pressures toward political systems that are frankly authoritarian by current standards"—see Ophuls,

Ecology and the Politics of Scarcity Revisited (New York: Freeman, 1992), p. 216. Theologian Thomas Berry expressed a similar fear: "The greatest danger to the human community may be the loss of its will to carry on the cosmic and numinous intentions within itself. The danger is the loss of internal vitality and a cooling down of life energies." See Berry, *Evening Thoughts*, pp. 7, 71, 136. Robert Heilbroner, "Second Thoughts on the Human Prospect," *Challenge* (May–June 1975): 27; also Peter Burnell, "Climate Change and Democratization" (Berlin: Heinrick Böll Stiftung, 2009), p. 91. Consider as well Cambridge University astronomer Martin Rees, who gave humanity a 50–50 chance of surviving to the year 2100; Rees, *Our Final Hour*; Giddens, *The Politics of Climate Change*; Robert Rothkopf, *Power Inc* (New York: Farrar, Straus and Giroux, 2012), p. 360.

26 David W. Orr and Stuart Hill, "Leviathan, the Open Society, and the Crisis of Ecology," *Western Political Quarterly* 31, no. 4 (December 1978): 457–469.

27 To arrive at such conclusions one must overlook the recurring instances of corporate boondoggles, corruption, malfeasance, and incompetence: remember ENRON, worldcom, Lehman Bros., corporate collusion in the banking collapse of 2008, as well as the standard-issue everyday law-breaking done in the name of increasing a market share or avoiding regulation. One need not demonize persons but should be alert to the pernicious effects traceable to the rules of the system that aim to maximize short-term shareholder value and little else. Amory B. Lovins et al., *Reinventing Fire* (White River Junction, VT: Chelsea Green, 2011), p. ix. David Coady et al., "How Large Are Global Energy Studies?," International Monetary Fund Working Paper, WP/15/105 (2015). For the history of energy subsidies in the United States, see Nancy Pfund and Ben Healy, *What Would Jefferson Do? The Historical Role of Federal Subsidies in Shaping America's Energy Future* (San Francisco: DBL Investors, 2011).

28 Steve Coll, *Private Empire: ExxonMobil and American Power* (New York: Penguin, 2012).

29 Karl Polanyi, *The Great Transformation* (1944; Boston: Beacon Press, 1967), p. 73; Michael Sandel, *What Money Can't Buy: The Moral Limits of Markets* (New York: Farrar, Straus and Giroux, 2012); Robert Kuttner, *Everything for Sale* (New York: Knopf, 1997).

30 Paul Hawken, *Blessed Unrest* (New York: Penguin, 2007); Steve Waddell, *Global Action Networks* (New York: Palgrave-Macmillan, 2011),

p. 23; Mazower, *Governing the World*, pp. 418, 420. See also the hymn to billionaires by Matthew Bishop and Michael Green: *Philanthropocapitalism* (New York: Bloomsbury, 2008).

31 Gary Gerstle, *Liberty and Coercion* (Princeton, NJ: Princeton University Press, 2015), pp. 1, 340, 350.

32 Dahl, *How Democratic Is the American Constitution?* Eric Nelson makes a roughly similar argument in *The Royalist Revolution* (Cambridge, MA: Harvard University Press, 2015); Myerson, "Foundering Fathers," p. 16.

33 Thomas Edsall, *The Age of Austerity* (New York: Doubleday, 2012), pp. 149, 185.

34 Naomi Klein, "Capitalism vs. the Climate," *The Nation* (November 21, 2011); see also her book *This Changes Everything* (New York: Simon & Schuster, 2014).

35 Benjamin Barber, *Strong Democracy* (Berkeley: University of California Press, 1984), pp. 117, 151; Also Michael Shuman, *Local Economy Solution*; Thad Williamson, David Imbroscio, and Gar Alperovitz, *Making a Place for Community* (New York: Routledge, 2002). All make a compelling case for grounding democracy in local ownership; see Barber, *Strong Democracy*, p. 269.

36 Amy Gutmann and Dennis Thompson, *Why Deliberative Democracy* (Princeton, NJ: Princeton University Press, 2004), pp. 7, 59. Political scientist James Fishkin has proposed establishing processes of deliberative polling, which would assemble random groups of citizens in a jury-like setting to weigh policy positions of candidates for public offices. See James S. Fishkin, *The Voice of the People: Public Opinion and Democracy* (New Haven: Yale University Press, 2004), p. 171; also Bruce Ackerman and James Fishkin, *Deliberation Day* (New Haven: Yale University Press, 2004), p. 171; Beth Simone Noveck, *Smart Citizens, Smarter State* (Cambridge, MA: Harvard University Press, 2015). Noveck's critique of democratic theory is important. She proposes to harness the power of the internet to engage citizen experts in actual policymaking.

37 Levinson, *Framed*, p. 389.

38 See John Keane's magisterial *The Life and Death of Democracy* (New York: Norton, 2009); Paul Woodruff, *First Democracy: The Challenge of an Ancient Idea* (New York: Oxford University Press, 2005); and John Plamenatz, *Democracy and Illusion* (London: Longman, 1973), p. 9.

39 Walter Prescott Webb, *The Great Frontier* (1951; Austin: University of Texas Press, 1964); David Potter, *People of Plenty* (Chicago: University of Chicago Press, 1954); see also Andrew Bacevich, *The Limits of Power*

(New York: Metropolitan Books, 2008), pp. 15–66; Robert Putnam, *Bowling Alone* (New York: Simon & Schuster, 2000); and Burnell, "Climate Change," p. 40. John Keane believes that democracies are "sleepwalking their way into deep trouble," Keane, *Life and Death of Democracy*, xxxii.

40 Thomas Mann and Norman Ornstein, *It's Even Worse Than It Looks* (New York: Basic Books, 2012); Theda Skocpol and Vanessa Williamson, *The Tea Party and the Remaking of Republican Conservatism* (New York: Oxford University Press, 2012); Jill Lapore, *The Whites of Their Eyes* (Princeton, NJ: Princeton University Press, 2010); Erich Fromm, *Beyond the Chains of Illusion* (1962; New York: Continuum, 1990), p. 119; Richard Hofstadter, *Anti-Intellectualism in American Life* (New York: Vintage Books, 1962); and Susan Jacoby, *The Age of American Unreason* (New York: Pantheon, 2008).

41 Richard Weaver, *Ideas Have Consequences* (1948; Chicago: University of Chicago Press, 1984); Jean Twenge and Keith Campbell, *The Narcissism Epidemic* (New York: Free Press, 2009), p. 276; Christopher Lasch, *The Culture of Narcissism* (New York: Norton, 1979); and Ian Mitroff and Warren Bennis, *The Unreality Industry* (New York: Oxford University Press, 1989), p. 21.

42 Mariana Mazzucato, in her book *The Entrepreneurial State* (London: Anthem Press, 2014), convincingly argues that the state typically has taken on risky early stage ventures when private capital would not or could not. In recent years, its role has been considerably and deliberately denigrated and the actual history wiped clean.

43 Jared Diamond, *Collapse: How Societies Choose to Fail or Succeed* (New York: Viking, 2005), p. 438. According to Diamond, societies collapse when they fail to anticipate a problem, fail to perceive it once it has arisen, fail to attempt to solve it after it has been perceived, or fail in the attempt to solve it. Some problems grow beyond the capacity of the society to solve, into what Thomas Homer-Dixon calls an "ingenuity gap." Consider also Harold Lasswell's classic definition of politics in *Politics: Who Gets What, When, and How* (1958; Cleveland: Meridian Books, 1968).

44 Dianne Dumonski, *The End of the Long Summer* (New York: Crown, 2009), p. 216.

45 See, for example, James Gustave Speth, *America the Possible* (New Haven: Yale University Press, 2012); Gar Alperovitz, *America beyond Capitalism* (Takoma Park, MD: Democracy Collaborative Press, 2011); Jeff

Gates, *Democracy at Risk* (Cambridge, Eng.: Perseus, 2000); Tim Jackson, *Prosperity without Growth* (London: Earthscan, 2009); Peter Victor, *Managing without Growth* (Northhampton, Eng.: Edward Elgar, 2008); Christopher Stone, *Should Trees Have Standing?* (Los Altos, CA: William Kaufman, 1972); also Cormac Cullinan, *Wild Law* (White River Junction, VT: Chelsea Green, 2011); and Thomas Berry, *Evening Thoughts* (San Francisco: Sierra Club Books, 2006), p. 44.

Chapter 7. Mind

Epigraphs. Epigraph to E. M. Forster, *Howards End* (London, 1910); G. K. Chesterson, *Orthodoxy* (New York: Image Books, 1959), p. 35.

1 Jenna Jambeck et al., "Plastic Waste Inputs from Land into the Ocean," *Science* 347, no. 6223 (February 13, 2015), pp. 768–771. By one study 90 percent of seabirds have plastic in their innards; see Chris Wilcox et al., "Threat of Plastic Pollution to Seabirds Is Global, Pervasive, and Increasing," *Proceedings of the National Academy of Sciences* (2015); available online at http://www.pnas.org/content/112/38/11899.full.pdf?sid=a3e10d85-cc05-4c2e-9100-62eddfec2230 (accessed February 25, 2016). Peter Kershaw, in *Biodegradable Plastics and Marine Litter* (Nairobi: UN Environmental Program, 2015), reports on the limited effectiveness of biodegradable plastics in solving the problem. The total weight of plastics could exceed the total weight of marine life in the world's oceans by 2050 according to a report from the World Economic Forum and the Ellen MacArthur Foundation, *The New Plastics Economy* (2016).

2 Annual Report of the President's Cancer Panel, *Reducing Environmental Cancer Risks* (Washington, DC, 2010), note 154.

3 There are other systems and processes with similar gyre-like properties, including the trillions of dollars that cycle through financial markets each day in search of higher rates of short-term profit. There are the petrochemicals that cycle through industrially managed agricultural ecologies and oceanic dead zones. The hallmarks in every case are ecological and human ruin and a foreshortening of the human horizon.

4 U. Thara Srinivasan et al., "The Debt of Nations and the Distribution of Ecological Impacts from Human Activities," *Proceedings of the National Academy of Sciences* 105, no. 5 (February 5, 2008): 1768–1773, note 166.

5 Jean-Pierre Dupuy, *The Mark of the Sacred* (Stanford, CA: Stanford University Press, 2013), p. 21; Mary Christina Wood, *Nature's Trust* (New York: Cambridge University Press, 2014).

6 Tom Lewis, *Divided Highways* (New York: Penguin Books, 1999), p. 133.

7 David Korten, *The Great Turning: From Empire to Earth Community* (San Francisco: Berrett Koehler, 2006); Hannah Arendt, *The Human Condition* (1958; Chicago: University of Chicago Press, 1970), p. 3; Susan Jacoby, *The Age of American Unreason* (New York: Pantheon, 2008), p. 307.

8 Richard Louv, *The Last Child in the Woods* (Chapel Hill, NC: Algonquin Books, 2005); Louv, *The Nature Principle* (Chapel Hill, NC: Algonquin Books, 2011), p. 142.

9 Erwin Chargaff, "Knowledge without Wisdom," *Harpers Magazine* (May 1980): 41–48; Wendell Berry, "Solving for Pattern," in *The Gift of Good Land* (San Francisco: Northpoint Press, 1981), pp. 134–145; Fritjof Capra and Pier Luigi Luisi, *The Systems View of Life* (Cambridge, Eng.: Cambridge University Press, 2014).

10 Edward LeRoy Long, Jr., *Higher Education as a Moral Enterprise* (Washington, DC: Georgetown University Press, 1992), p. 220.

11 David W. Orr, *Earth in Mind* (1994; Washington, DC: Island Press, 2004); *Ecological Literacy* (Albany: State University of New York Press, 1992); Michael Crow, "None Dare Call It Hubris: The Limits of Knowledge," *Issues in Science and Technology* (Winter 2007).

12 C. A. Bowers, *Educating for an Ecologically Sustainable Culture* (Albany: State University of New York Press, 1995); C. A. Bowers, *Education, Cultural Myths, and the Ecological Crisis* (New York: State University of New York Press, 1993).

13 Eileen Crist, "On the Poverty of Our Nomenclature," *Environmental Humanities* 3 (2013): 137; Crow, "None Dare Call It Hubris"; Robert Sinsheimer, "The Presumptions of Science," *Daedalus* 107, no. 2 (Spring 1978): 23–35; Roger Shattuck, *Forbidden Knowledge* (New York: St. Martin's Press, 1996). Hannah Arendt wrote that these are "political question[s] of the first order and therefore can hardly be left to the decision of professional scientists or professional politicians"; see Arendt, *The Human Condition* (Chicago: University of Chicago Press, 1958), p. 3.

14 James Lovelock, "A Book for All Seasons," *Science* 280 (May 8, 1998): 832–833.

15 See the special January 2014 issue of *Solutions*, which is devoted to the Oberlin Project, as well as Chapter 12.

16 Fritjof Capra and Pier Luigi Luisi, *The Systems View of Life*. See C. A. Bowers on ecological intelligence; Bowers, *University Reform in an Era of Global Warming* (Eugene: Eco-justice Press, 2011), pp. 155–188.

17 Peter Senge, *The Fifth Discipline* (1990; New York: Doubleday, 2006).

18 Matthew D. Lieberman writes, "I see a tapestry of neural systems woven together to bind us to one another." See Lieberman, *Social: Why Our Brains Are Wired to Connect* (New York: Crown, 2013), p. 302.

19 Janine Benyus, lecture at the Omega Institute, September 6, 2013; Michael Pollan, "The Intelligent Plant," *New Yorker* (December 23 and 30, 2013): 92–105; Stefano Mancuso and Alessandra Viola, *Brilliant Green* (Washington, DC: Island Press, 2015).

20 Teilhard de Chardin, *The Phenomenon of Man* (New York: Harper Torchbook, 1965), pp. 180–184.

21 Forster, *Howards End*, p. 248.

22 Iain McGilchrist, *The Master and His Emissary* (New Haven: Yale University Press, 2010) is a brilliant discussion of baffling complexities of the mind and the brain and the possibility that they are not of one mind—being either in competition or even at war; and what if there is some part of us that survives our death, often called a soul. In the face of such vaporous but perhaps real possibilities beyond our science, Marilynne Robinson recommends, with John Locke, that "we must 'sometimes be content to be very ignorant.'" See Robinson, *The Givenness of Things* (New York: Farrar, Straus and Giroux, 2015), pp. 14–15, 189.

23 Bill Vitek and Wes Jackson, eds., *The Virtues of Ignorance* (Lexington: University of Kentucky Press, 2008).

24 Mary Midgley, "Why Smartness Is Not Enough," in Mary Clark and Sandra Wawrytko, eds., *Rethinking the Curriculum* (Westport, CT: Greenwood Press, 1990), pp. 39–52.

Chapter 8. Heart

Epigraphs. Henry David Thoreau, *Walden* (Princeton, NJ: Princeton University Press, 2004), chapter 1; Czesław Miłosz, *The Separate Notebooks* (New York: Ecco, 1986), n.p.

1 Paul L. Montgomery, "A Failed Giveaway Meets a Foiled Getaway," *New York Times*, March 13, 1982.

2 Mark Dowie, *American Foundations: An Investigative History* (Cambridge, MA: MIT Press, 2002), pp. 106–140.

3 Diane Ravitch, *The Death and Life of the Great American School System* (New York: Basic Books, 2010), pp. 194–222; and Ravitch, *Reign of Error* (New York: Knopf, 2013), pp. 10–43. Linsey McGoey, *No Such Thing as a Free Gift: The Gates Foundation and the Price of Philanthropy* (New York:

Verso, 2015) documents the ironies of philanthrocapitalism and dubious record of the Gates Foundation in particular; see also Dale Russakoff, *The Prize* (Boston: Houghton Mifflin Harcourt, 2014). The Zuckerbergs have pledged to give away 99 percent of their Facebook stock, but apparently to a limited liability corporation that they will control and that will lower their tax burden. See Cynthia Powers, "Corporate Charity Is Corporate Power," www.truthout.org (December 24, 2015).

4 Margaret Atwood, *Payback* (Toronto: Anansi, 2008), p. 179; Mary Douglas, foreword to Marcel Mauss, *The Gift* (New York: Norton, 2000), pp. vii–viii. How this translates into public policies such as Social Security and health insurance created by legislation and funded by taxation remained an unsolved problem according to Douglas and Marcel Mauss; see Mauss, *The Gift*, p. 69; Lewis Hyde, *The Gift* (New York: Vintage, 1983), pp. 19, 23, 89, 139.

5 William Shakespeare, *Hamlet*, 3.1.101.

6 Hyde, *The Gift*, p. 89.

7 Mauss, *The Gift*, p. 72.

8 The National Philanthropic Trust, *Giving USA, 2015* (June 2015). Individuals reportedly gave $258 billion and corporations $17 billion—Robert Reich estimates that the lost revenue to the U.S. Treasury is $40 billion per year. As foundations multiplied and grew, a new species of human is said to have evolved: the "philanthropoid" or program officer. They are amazingly intelligent, even wise, and their jokes are very funny. For these reasons, admiring supplicants eagerly return their emails and phone calls and avidly seek their advice on all manner of things.

9 Joel Fleishman, *The Foundation* (New York: Public Affairs, 2007), pp. xv, 150–152.

10 Larry Kramer and Carol Larsen, presidents of the Hewlett and Packard foundations, respectively, have described climate change as an "everything problem . . . with the unique potential to undermine everything we care about as foundations." Kramer and Larsen, "Foundations Must Move Fast to Fight Climate Change," *Chronicle of Philanthropy* (April 20, 2015); Gara Lamarche, "Democracy and the Donor Class," *Democracy: A Journal of Ideas* 34 (Fall 2014): 56–57.

11 Or in Naomi Klein's words, "The fetish of centrism—of reasonableness, seriousness, splitting the difference, and generally not getting overly excited about anything. This is the habit of thought that truly rules our era." Naomi Klein, *This Changes Everything* (New York: Simon & Schuster, 2014), p. 22.

12 David W. Orr, "Cleveland in a Hotter Time and What the Cleveland Foundation Can Do About It," report to the Cleveland Foundation, May 2015.

13 Rob Reich, *Boston Review*, March 1, 2013.

14 Encyclical Letter, *Laudato Si'*, June 18, 2015, p. 17.

15 Seneca, De Beneficiis, *Seneca: Moral Essays*, vol. 3, trans. John Basore (Cambridge, MA: Harvard University Press, Loeb Library, 1935); Cicero, *On Obligations*, trans. P. G. Walsh (Oxford, Eng.: Oxford University Press, 2001); Dante Alighieri, *The Divine Comedy*, trans. Allen Mandelbaum (New York: Alfred Knopf, 1995), *Inferno*, pp. xxxiv, 209–213; Margaret Visser, *The Gift of Thanks* (New York: HarperCollins, 2008), p. 311; Johann Wolfgang von Goethe, *Maxims and Reflections* (London: Penguin, 1998), p. 21.

16 Visser, *Gift of Thanks*, pp. 321, 322.

17 Aldo Leopold, "The Land Ethic," in Curt Meine, ed., *Aldo Leopold* (New York: Library of America, 2013), p. 189; Atwood, *Payback*, pp. 201–202.

18 Atwood, *Payback*, 203; Dale Jamieson, *Reason in a Dark Time: Why the Struggle against Climate Change Failed—and What It Means for Our Future* (Oxford, Eng.: Oxford University Press, 2014), p. 8; Visser, *Gift of Thanks*, p. 392.

19 Letter to James Madison, September 6, 1789, in Merrill Peterson, *Jefferson* (New York: New American Library, 1984), pp. 959–964, which went on to remark that "no man can by *natural right* oblige the lands he occupied, or the persons who succeed him in that occupation, to the payment of debts contracted by him . . . the earth belongs always to the living generation. They may manage it then, and what proceeds from it, as they please, during their usufruct . . . this principle that the earth belongs to the living and not to the dead."

20 Edmund Burke, *Reflections on the Revolution in France* (1790; London: Penguin Books, 1986), p. 119; ibid., 195; Richard Bourke, *Empire & Revolution* (Princeton, NJ: Princeton University Press, 2015), pp. 677–688.

21 John Locke, *Two Treatises of Government*, ed. Peter Laslett (1688; New York: Mentor Books, 1965), pp. 329, 332.

22 Burke, *Reflections on the Revolution in France*, p. 119.

23 Stephen Gardiner, *A Perfect Moral Storm* (New York: Oxford University Press, 2011) is a brilliant analysis of, among other things, the complexities of tradeoffs between generations.

24 A. P. d'Entrèves, *Natural Law* (London: Hutchinson University Library, 1957), pp. 33–35; Leszek Kołakowski, *Is God Happy?* (New York: Basic Books, 2013), pp. 241–250.

25 Leopold, "Land Ethic," p. 172; Aldo Leopold, *Round River*, ed. Luna Leopold (New York: Oxford University Press, 1972), p. 64.

26 Visser, *Gift of Thanks*, p. 386; Frans De Wall, *The Bonobo and the Atheist* (New York: Norton, 2013).

27 Abraham Heschel, *Man Is Not Alone: A Philosophy of Religion* (New York: Farrar, Straus and Giroux, 1951), p. 37.

Chapter 9. The Long Revolution

Epigraphs. Reinhold Niebuhr, *The Irony of American History* (New York: Charles Scribner's Sons, 1952), p. 63; Martin Amis, *The War against Cliché* (London: Vintage, 2002), p. 306.

1 Amory Lovins, "Energy Strategy: The Road Less Traveled," *Foreign Affairs* (October 1976); Amory Lovins, *Soft Energy Paths* (New York: Penguin, 1977); Gerald O. Barney, ed., *The Global 2000 Report* (Washington: U.S. Government Printing Office, 1980); Dennis Meadows et al., *The Limits to Growth* (New York: Universe Books, 1972); Editors of *The Ecologist, Blueprint for Survival* (Boston: Houghton Mifflin, 1972); The Union of Concerned Scientists, "World Scientists' Warning to Humanity" (Cambridge, MA: Union of Concerned Scientists, 1992); Gro Harlem Brundtland, *Our Common Future* (New York: Oxford University Press, 1987), and U.N. Conferences in Stockholm in 1972 and Rio de Janeiro in 1992. It is entirely plausible that CO_2 emissions could have been capped below 400 ppm and that the events of September 11, 2001, and the aftermath of the ill-conceived "War on Terror" would not have happened. The more important issues are, however, whether we can still learn from colossal errors of judgment, policy, and politics; see Kevin Coyle and Lise Van Susteren, *The Psychological Effects of Global Warming* (Washington, DC: National Wildlife Federation, 2012).

2 Henry Adams to Charles Francis Adams, April 11, 1862, in J. C. Levenson et al., eds., *The Letters of Henry Adams, 1858–1868* (Cambridge, MA: Harvard University Press, 1982), p. 290.

3 Henry Adams, *The Education of Henry Adams* (New York: Library of America, 1983); Siegfried Giedion, *Mechanization Takes Command* (1948; New York: Norton, 1969), p. 716; Lewis Mumford, "The Uprising of Caliban," in Mumford, *Interpretations and Forecasts: 1922–1972*

(New York: Harcourt Brace Jovanovich, 1973), p. 349; Václav Havel, *Living in Truth* (London: Faber and Faber, 1986), pp. 146, 159.

4 Pesticides that originated in wartime chemical labs, for example, were then applied in a war on bugs; Edmund Russell, *War and Nature* (Cambridge, Eng.: Cambridge University Press, 2001). Lewis Mumford compared our situation to Goethe's tale of the Sorcerer's Apprentice who forgot "the Master Magician's formula for regulating or halting the flood and so are on the point of drowning in it." Mumford, "Bacon: Science as Technology," in Mumford, *Interpretations and Forecasts*, p. 166; Richard Sclove, *Democracy and Technology* (New York: Guilford Press, 1995); Nichols Fox, *Against the Machine* (Washington, DC: Island Press, 2002), pp. 285–329; Jill Lepore, "The Disruption Machine," *New Yorker* (June 23, 2014): 30–36. The history is important but mostly neglected. See Kirkpatrick Sale, *Rebels against the Future* (Reading: Addison-Wesley, 1995); Neil Postman, *Technopoly* (New York: Knopf, 1992); David Ehrenfeld, *The Arrogance of Humanism* (New York: Oxford, 1978); and Jacques Ellul, *The Technological Society* (New York: Vintage Books, 1964). Brian Arthur's *The Nature of Technology* (New York: Free Press, 2009) is a thoughtful analysis of "what technology is and how it comes to be," concluding with the view that it is good when it extends "our nature," bad when it "enslaves" (p. 216). Exactly how we distinguish between these extremes in time to avoid enslavement is not discussed, maybe because we have no means by which to decide in advance. Kevin Kelly, in *What Technology Wants* (New York: Viking, 2010) ranges between the euphoric: "Look what is coming: stitching together all the minds of the living, wrapping the planet in a vibrating cloak of electronic nerves, entire continents of machines conversing with one another . . . How can this not stir that organ in us that is sensitive to something larger than ourselves?" (p. 358), and worry: "humans are both master and slave to the technium, and our fate is to remain in this uncomfortable dual role" (p. 187); "technology chips away at our human dignity" (p. 197); and "accelerated cycles of creation can race so far ahead of our intentions that it is worrisome" (p. 259). Kelly is conflicted, caught between fascination and the fear that the juggernaut of technology is beyond any human control. All in all, a reasonable place to be stuck. Ed Ayres, *Defying Dystopia* (New Brunswick: Transaction Press, 2016), is a sweeping and thoughtful critique of pervasive technology and its counterintuitive effects.

5 Edward O. Wilson, *Half-Earth: Our Planet's Fight for Life* (New York: Liveright Publishing, 2016), p. 211.

6 National Oceanic and Atmospheric Administration, *State of the Climate in 2014*, special supplement to the *Bulletin of the American Meteorological Society* 96, no. 7 (July 2015); the classic study is William L. Thomas, Jr., ed., *Man's Role in Changing the Face of the Earth* (Chicago: University of Chicago Press, 1956); and a subsequent volume, B. L. Turner II et al., eds., *The Earth as Transformed by Human Action* (Cambridge, Eng.: Cambridge University Press, 1990); Bill McKibben, *Eaarth* (New York: Times Books, 2010).

7 This is not happening rapidly enough, however, according to the International Energy Agency, *Energy and Climate Change* (Paris: IEA, 2015), which states, "For the first time since the I.E.A. started monitoring clean energy progress, not one of the technology fields traced is meeting its objectives" relative to climate goals. The United States, in particular, lags in energy research despite recent progress. See Eduardo Porter, "Innovation Sputters in Battle against Carbon," *New York Times*, July 22, 2015; Al Gore, "The Turning Point," *Rolling Stone* (July 3–17, 2014): 93. His comment, however, does not consider the ongoing effects of heat-trapping gases in the atmosphere, which is not presently a solvable problem.

8 The pace of technological changes continues to accelerate. See U.S. Department of Energy, *Revolution . . . Now: The Future Arrives for Five Clean Energy Technologies, 2015 Update* (Washington: USDOE, November, 2015); www.nextgenclimate.org; Ken Zweibel et al., "A Solar Grand Plan," *Scientific American* (January 2008): 64–73; Amory Lovins et al., *Winning the Oil Endgame* (Snowmass, CO: Rocky Mountain Institute, 2004); Greenpeace, *Energy [R]evolution: A Sustainable World Energy Outlook* (2015). A network of cities has formed to reduce carbon emissions as reported in C-40 Cities and ARUP, *Climate Action in Megacities 3.0* (2015). The Compact of Mayors led by Michael Bloomberg includes 360 cities worldwide with similar intentions of making up in carbon reductions what national pledges fail to do. The Carbon Neutral Cities Alliance of seventeen global cities aims for 80 percent reductions in CO_2 emissions by 2050.

9 James Hansen et al., "Ice Melt, Sea Level Rise and Superstorms," *Atmospheric Chemistry and Physics Discussions* (2015); Jonathon Porritt, *The World We Made* (London: Phaidon, 2014) is a picture of what is possible with the proper application of ingenuity, design, and technology. James Kuntsler is not so confident: see his *World Made by Hand* (New York: Grove Press, 2008); Jared Diamond, *Collapse* (New York: Viking, 2005); Joseph Tainter, *The Collapse of Complex Societies* (Cambridge, Eng.: Cambridge

University Press, 1988); Ronald Wright, *A Short History of Progress* (New York: Carroll & Graf, 2004); and Margaret Atwood, *Oryx and Crake* (New York: Anchor Books, 2004), among others.

10 Encyclical Letter, *Laudato Si'*, June 18, 2015, p. 104.

11 Leopold quoted in Curt Meine, ed., *Leopold: A Sand County Almanac and Other Writings on Ecology and Conservation* (New York: Library of America, 2013), p. 177; Václav Havel, *Summer Meditations* (New York: Knopf, 1992), p. 116 (emphasis added).

12 Cynthia Moe-Lobeda, *Resisting Structural Evil* (Minneapolis: Fortress Press, 2013), p. 198.

13 Erich Fromm, *To Have or to Be* (New York: Bantam Books, 1981).

14 Viktor Frankl, *Man's Search for Meaning* (New York: Simon & Schuster, 1984), p. 96.

15 Ibid., p. 100.

16 Thomas Berry, *Evening Thoughts* (San Francisco: Sierra Club Books, 2006), p. 43; Historian of religion Mircea Eliade's distinction between sacred and profane attitudes to the world echo through Berry's writings. See Eliade, *The Sacred and the Profane* (New York: Harcourt Brace Jovanovich, 1957), pp. 201–205; Thomas Berry, *The Great Work* (New York: Bell Tower, 1999), p. 4. Berry's thought coincides with that of Louv and others: "For children to live only in contact with concrete and steel and wires and wheels and machines and computers and plastics, to seldom experience any primordial reality or even to see the stars at night, is a soul deprivation that diminishes the deepest of their human experience." Berry, *The Great Work*, p. 82.

17 Brian Thomas Swimme and Mary Evelyn Tucker, *Journey of the Universe* (New Haven: Yale University Press, 2011), p. 66; also Brian Swimme and Thomas Berry, *The Universe Story* (New York: Harper Collins, 1992).

18 Swimme and Tucker, *Journey of the Universe*, pp. 112–113; John Grim and Mary Evelyn Tucker, *Ecology and Religion* (Washington, DC: Island Press, 2014) is a superb recalibration of religion in nature, leading us to "see nature not simply as a resource but as the source of life—for humans and the entire Earth community" (p. 170).

19 E. O. Wilson, *Biophilia* (Cambridge, MA: Harvard University Press, 1982); Stephen Kellert, *Building for Life* (Washington, DC: Island Press, 2005).

20 Richard Louv's *The Nature Principle* (Chapel Hill, NC: Algonquin Books, 2012) is a brilliant summary of the evidence and its importance

for human flourishing; so, too, is Wallace Nichols, *Blue Mind* (New York: Back Bay Books, 2014); and Atul Gawande, *Being Mortal* (New York: Metropolitan Books, 2014). See also Kellert, *Building for Life*; Stephen Kellert and Judith Heerwagen, *Biophilic Design* (New York: John Wiley & Sons, 2008); and Timothy Beatley, *Biophilic Cities* (Washington, DC: Island Press, 2011).

21 Elizabeth Kolbert, *The Sixth Extinction* (New York: Henry Holt, 2014); Gerardo Ceballos et al., "Accelerated Modern Human-Induced Species Losses: Entering the Sixth Mass Extinction," *Science* (June 19, 2015); Kevin Coyle and Lise Van Susteren, *The Psychological Effects of Global Warming* (Washington, DC: National Wildlife Federation, 2012). Ernest Becker's *Denial of Death* (New York: Free Press, 1974) launched the conversation about what has been named by psychologists "terror management." The notable difference is that Becker's concern was to redirect Sigmund Freud's theory to our own mortality, not to the dying of nature all around us.

22 George Orwell, *A Collection of Essays* (1946; New York: Harcourt Brace Jovanovich, 1981), p. 120; Encyclical Letter, *Laudato Si'*, p. 40.

23 Encyclical Letter, *Laudato Si'*, p. 67.

24 Wolfgang Sachs et al., *Greening the North* (London: ZED Books, 1998); Wolfgang Sachs, ed., *The Development Dictionary* (London: ZED Books, 1992).

25 Fritjof Capra and Pier Luigi Luisi, *The Systems View of Life: A Unifying Vision* (Cambridge, Eng.: Cambridge University Press, 2014); See also Fritjof Capra, *The Web of Life* (New York: Anchor Books, 1996). The Indo-European root word is *Kailo*. György Doczi, *The Power of Limits* (Boston: Shambhala: 1981), p. 133; Alan Savory, *Holistic Management*, 2nd ed. (Washington, DC: Island Press, 1999).

26 See Paul Hawken's Project Drawdown at www.drawdown.org. According to Wes Jackson at the Land Institute, estimates of the potential for reforestation and better management of farm and rangeland to remove carbon from the atmosphere vary by an order of magnitude. The analysis is further complicated by factoring in changing rainfall patterns and rising temperatures that will affect the capacity of soils and biomass to accumulate carbon. Further, the net carbon stored must subtract that released in the making, transporting, and dispersing of soil amendments like biochar. And then there are the complications of securing compliance at a massive scale across millions of acres here and elsewhere; see George Monbiot, *Feral* (London: Penguin Books, 2013), p. 6.

27 See Wallace Stegner's classic statement for wilderness, "Wilderness Letter" (1960), in Page Stegner, ed., *Marking the Sparrow's Fall* (New York: Henry Holt, 1998), pp. 111–120; Howie Wolke, "Wilderness: What and Why?" in George Wuerthner et al., eds., *Keeping the Wild* (Washington, DC: Island Press, 2014).

28 Monbiot, *Feral*, p. 167.

29 Edward O. Wilson, *Half Earth* (New York: Liveright, 2016).

30 From Dave Foreman with the Rewilding Institute and Wildlands Network. See also Deborah Popper and Frank Popper, "From Dust to Dust," *Planning* (1987).

31 Václav Havel, *The Art of the Impossible* (New York: Knopf, 1997), p. 91; Stuart Brand, *Whole Earth Discipline* (New York: Penguin, 2009); and Mark Lynas, *The God Species* (Washington, DC: National Geographic, 2011); Encyclical Letter, *Laudato Si'*, p. 43. "An EcoModernist Manifesto" is a mixture of common sense, obvious, and the absurd—more "modernist" than "eco." The authors have made their peace with nuclear power, genetic engineering, nanotechnologies, and so on, as a way to "decouple" humans from dependence on natural systems.

32 Stuart Russell, a pioneer in the field of artificial intelligence, says that "to those who say, well, we may never get to human-level or superintelligent AI, I would reply: It's like driving straight toward a cliff and saying, 'let's hope I run out of gas soon!'" *Science* 349, no. 6245 (July 17, 2015): 252. The warnings are in fact as ancient as the myth of Pandora. In their more recent form, Mary Shelley's *Frankenstein* or Melville's *Moby Dick*, the themes include the lack of responsibility and obsession. Much of the same applies to the synthetic biology movement, which aims to create a parallel world of engineered biological organisms. Having botched the management of the standard flora and fauna of Earth, why would we do a better job with a new and probably intrusive and destructive biology? And how would the original biota and the conjured get along? For more, see Laurie Garrett, "Biology's Brave New World," *Foreign Affairs* 92, no. 6 (November–December 2013); Richard Lewontin's response "The New Synthetic Biology: Who Gains?" *New York Review of Books* (May 8, 2014); as well as Bill Joy, "Why the Future Doesn't Need Us," *Wired* (April 2000); Nick Bostrom, *Superintelligence* (Oxford, Eng.: Oxford University Press, 2014); and James Barrat, *Our Final Invention* (New York: Thomas Dunne Books, 2013). "Autonomous Weapons: An Open Letter from AI & Robotics Researchers," Future of Life Institute, July 28, 2015. The

letter is signed by Stephen Hawking, Elon Musk, Daniel Dennett, and twelve thousand others; Barrat reports that fifty-six countries already have or are developing battlefield robots. See Barrat, *Our Final Invention*, p. 21; "Human Rights Watch," op-ed in the *Boston Globe*, September 15, 2015. Martin Ford, in his *The Rise of the Robots* (New York: Basic Books, 2015), proposes a guaranteed income for the unemployed without explaining either how humans would adapt to being rather useless, unskilled, inept, and perpetually bored or who would finance a guaranteed income for millions.

33 At the far fringe of organized gullibility, some even believe that Fox News has something to do with objective reality. But the problem has deep roots. See Richard Hofstadter, *Anti-Intellectualism in American Life* (New York: Vintage, 1963); and Susan Jacoby, *The Age of American Unreason* (New York: Pantheon, 2008).

34 Gregory Bateson, *Mind and Nature: A Necessary Unity* (New York: Dutton, 1979); and Bateson, *Steps to an Ecology of Mind* (New York: Ballantine, 1975). "Wisdom," he writes, is "the knowledge of the larger interactive system—that system which, if disturbed, is likely to generate exponential curves of change" (p. 433); and Donella Meadows, *Thinking in Systems* (White River Junction, VT: Chelsea Green, 2008).

35 Thomas Berry, *The Great Work* (New York: Bell Tower, 1999), p. 73.

36 C. S. Lewis, *The Abolition of Man* (1947; New York: MacMillan, 1965), pp. 79–80.

37 Michael Crow and William Dabars, *Designing the New American University* (Baltimore: Johns Hopkins University Press, 2015), p. 233.

38 Crow and Dabars say it more soothingly: "Our universities remain disproportionately focused on perpetuating disciplinary boundaries and developing increasingly specialized new knowledge at the expense of collaborative endeavors targeting real-world problems" (ibid., 234). In other words, they settle for "busywork on a vast, almost incomprehensible scale," in historian Page Smith's words. Gregory Bateson once asked the California Board of Regents whether they "foster whatever will promote in students, in faculty, and around the boardroom table those wider perspectives which will bring our system back into an appropriate synchrony or harmony between rigor and imagination? As teachers, are we wise?" It was regarded as a rhetorical question. See Gregory Bateson, *Mind and Nature: A Necessary Unity* (New York: Dutton, 1979), p. 223.

39 William Deresiewicz, *Excellent Sheep: The Miseducation of the American Elite* (New York: Free Press, 2014).

40 Thomas Berry, *Great Work*, quotation on p. 85. See also James Farrell, "Good Work and the Good Life," in Kaethe Schwehn and L. DeAne Lagerquist, eds., *Claiming Our Callings* (New York: Oxford University Press, 2014); and Farrell, *The Nature of College* (Minneapolis: Milkweed Press, 2010).

41 Crow and Dabars, *Designing the New American University*, p. 242; See also Ernest L. Boyer, "Creating the New American College," *Chronicle of Higher Education* (March 9, 1994), p. A48.

42 For more on the Adam Joseph Lewis Center, see David W. Orr, *Design on the Edge* (Cambridge, MA: MIT Press, 2006). At the secondary level, Rachel Gutter and her colleagues at the Center for Green Schools are changing the standards for school facilities, which are, in turn, changing the curriculum as well.

43 Nicholas Maxwell, *From Knowledge to Wisdom* (London: Pentire Press, 2007); and Chet Bowers, *University Reform in an Era of Global Warming* (Eugene, OR: Eco-Justice Press, 2011). In one exemplary venture, Göran Broman and colleagues "call for the development of a new kind or next-generation of science, a systematic research approach linking transformative theory with enabling practice across the diversity of civilization's intellectual and functional pursuits, and which uses sustainability as a guide." See Broman et al., "Systematic Leadership towards Sustainability," *Journal of Cleaner Production* 64, nos. 1–2 (2014).

44 Samuel Huntington argues that order must come before democracy in his *Political Order in Changing Societies* (New Haven: Yale University Press, 1968); see also Frances Fukuyama, *Political Order and Political Decay* (New York: Farrar, Straus, and Giroux, 2014), particularly pp. 524–548.

45 Paul Light, *A Government Ill Executed* (Cambridge, MA: Harvard University Press, 2008) is an insightful analysis of how to reverse the decline in the federal service that harkens back to Alexander Hamilton and Thomas Jefferson's competing visions of the scope and scale of government.

46 Lynton Keith Caldwell, *The National Environmental Policy Act* (Bloomington: Indiana University Press, 1998), is definitive on the creation, substance, and evolution of the act by the man who was its author.

47 Frank Kalinowski, "The Environmental Legacy of James Madison," in *Environmental Legacies* (New York: Palgrave-Macmillan, 2016).

48 Christopher Stone, *Should Trees Have Standing?* (Los Altos, CA: William Kaufmann, 1974); Mary Christina Wood's *Nature's Trust* (New York: Cambridge University Press, 2014) is a landmark in the dialogue about the rights of future generations. Better late than never, the legal community is bestirring itself to respond driven by suits initiated by groups such as Our Children's Trust in the United States and the Urgenda Foundation in Amsterdam. The basic idea is that injustice and the effects of criminal neglect spill across generational boundaries but can only be remedied before, not after, the fact. That is to say that the guilty party(ies) cannot be effectively summoned from the grave to the bar of justice, nor can the facts of diminished ecologies be reversed; see NextGen Climate, *The Economic Case for Clean Energy* (2015). Worldwide costs of weather-related disasters are estimated to be $250–300 billion per year and rising; see Center for Research on the Epidemiology of Disasters, *The Human Cost of Weather Related Disasters* (Brussels: UNISDR, 2015), p. 5. Tim Mulgan, *Ethics for a Broken World* (Montreal: McGill-Queens University Press, 2011) is a provocative discussion of how future generations living in a broken world will regard us. The fact is that we have not "enfranchised future people or offered effective constitutional protection for future rights" (p. 220). Unless we act soon and effectively, they will regard us as a hardhearted, and perhaps a stupid, lot.

49 Michael Pollan, "The Intelligent Plant," *New Yorker* (December 23, 2013). Stefano Mancuso and Alessandra Viola, *Brilliant Green* (Washington, DC: Island Press, 2015) is a noteworthy summary of the evidence of plant intelligence; Carl Safina, *Beyond Words* (New York: Henry Holt, 2015) is a moving account of animal intelligence and prescience. Taken together, one might conclude that Cartesian views of intelligence just weren't all that intelligent. Marc Bekoff, *The Emotional Lives of Animals* (Novato, CA: New World Library, 2007); and Bekoff, *The Animal Manifesto* (Novato, CA: New World Library, 2010) document the richness of animal emotions, extending the work of Donald Griffin, *Animal Minds* (Chicago: University of Chicago, 1992); and Roger Fouts's wonderful *Next of Kin* (New York: Morrow, 1997); Irene Pepperberg, *Alex and Me* (New York: Harper, 2008); and Benedict Carey, "Brainey Parrot Dies Emotional to the End," *New York Times*, September 11, 2007, p. A21. Alex, an African gray parrot, could form rudimentary sentences; we haven't learned to do the same in parrot language. Marilynne Robinson writes of science that it "assert[s] that everything is explicable, that whatever has not been explained will be explained—and, furthermore, by

their methods. They have seen to the heart of it all. So mystery is banished—mystery being no more than whatever their methods cannot capture yet." See Robinson, *The Givenness of Things* (New York: Farrar, Straus, and Giroux, 2015), p. 14; Václav Havel, *Living in Truth* (London: Faber and Faber, 1986).

50 For example, Holmes Rolston, *Philosophy Gone Wild* (Buffalo: Prometheus Books, 1986); Rolston, *Environmental Ethics* (Philadelphia: Temple University Press, 1988); and a thoughtful reconsideration, Ben Minteer, *Refounding Environmental Ethics* (Philadelphia: Temple University Press, 2012). Robert Costanza et al., "The Value of the World's Ecosystem Services and Natural Capital," *Nature* 387 (May 1997): 253–260, updated in Costanza et al., "Changes in the Global Value of Ecosystem Services," *Global Change* 26 (2014). Global ecosystem services grew from an estimated $33 trillion in the 1997 study to $145 trillion in the 2014 article.

51 Peter Brown, *The Commonwealth of Life* (Montreal: Black Rose Books, 2007), p. 89; see also Brown, *Restoring the Public Trust* (Boston: Beacon Press, 1994), pp. 71–91; and Mary Christina Wood, *Nature's Trust* (Cambridge, Eng.: Cambridge University Press, 2014), pp. 125–142, 143, 150.

52 Mariana Mazzucato, *The Entrepreneurial State* (London: Anthem Press, 2013), pp. 62, 189.

53 Food scarcity will likely become endemic. See, for example, U.S. Global Change Program, *Climate Change and Global Food Security and the U.S. Food System* (Washington, DC: U.S. Global Change Program, 2015); Joel K. Bourne, Jr., *The End of Plenty* (New York: Norton, 2015); Lloyds of London, *Food System Shock: The Insurance Impacts of Acute Disruption to the Global Food Supply*, emerging risk report (2015); and Nafeez Ahmed, *Scientific Model Supported by UK Government Taskforce Flags Risk of Civilization's Collapse by 2040* (London: Insurge Intelligence, 2015). For the quotations see Stan Cox, *Any Way You Slice It* (New York: New Press, 2013), pp. 14, 241. See also Michael Klare, *The Race for What's Left* (New York: Metropolitan Books, 2012). Fat chance, according to Robert Paarlberg in *The United States of Excess* (New York: Oxford University Press, 2015). He finds our inability to respond to climate change rooted in "flaws in national character" (p. 190). See also Paul Roberts, *The Impulse Society* (New York: Bloomsbury, 2014), which places the flaw in well-honed narcissism and the impulse to instant gratification; as well as Peter Victor, *Managing without Growth* (Cheltenham,

Eng.: Edward Elgar, 2008); Tim Jackson, *Prosperity without Growth* (London: Earthscan, 2009); and Juliet Schor, *Plenitude* (New York: Penguin Press, 2010).

54 Andro Linklater, *Owning the Earth* (New York: Bloomsbury, 2013), pp. 93–108, 388, 397. He writes that "communal systems were almost defenseless against the individualized challenge of private property. The ability of paper-based structures of private property to harness the legal resources of an entire society, to direct them at a particular parcel of ground, and to offer financial rewards for success simply overwhelmed the oral, local, and conservative systems of communal land ownership that stood in their way" (p. 105). Elinor Ostrom, *Governing the Commons* (Cambridge, Eng.: Cambridge University Press, 1990).

55 Burns Weston and David Bollier, *Green Governance* (Cambridge, Eng.: Cambridge University Press, 2013), pp. xxv, 20, 189, 201–202; David Bollier, *Think Like a Commoner* (Gabriola Island, BC: New Society Publishers, 2014). See also Peter Barnes, *Capitalism 3.0* (San Francisco: Berrett-Koehler, 2006), pp. 135–153. No commons can endure without curtailing the rights of corporations to ride roughshod over those of communities. See Community Environmental Legal Defense Fund, *On Community Civil Disobedience in the Name of Sustainability* (Oakland, CA: PM Press, 2015).

56 Barnes, *Capitalism 3.0*, pp. 143–144; Peter Barnes, *Who Owns the Sky?* (Washington, DC: Island Press, 2001), p. 131; see also Peter Barnes et al., "Creating an Earth Atmospheric Trust," *Science* 319 (2008): 724; Robert Costanza et al., "Creating an Atmospheric Trust to Help Implement the Paris Climate Deal," Letter to *Nature* (December 2015).

57 Weston and Bollier, *Green Governance*, pp. 262–263.

58 Elinor Ostrom, "A Polycentric Approach for Coping with Climate Change," *Annals of Economics and Finance* 15, no. 1 (2014): 97; also Andrew Jordan et al., "Emergence of Polycentric Climate Governance and Its Future Prospects," *Nature Climate Change* 5, no. 11 (November 2015): 977–982; and Jessica Green, *Rethinking Private Authority* (Princeton, NJ: Princeton University Press, 2013).

59 Michael Vandenbergh, "Private Environmental Governance," *Cornell Law Review* 99, no. 1 (2013): 138. Private environmental governance, he writes, includes "actions . . . that are designed to achieve traditionally governmental ends such as managing the exploitation of common pool resources, increasing the provision of public goods, reducing

environmental externalities, or more justly distributing environmental amenities [and] include traditional standard-setting, implementation, monitoring, enforcement, and adjudication" (p. 146); see also Jonathan Cannon, *Environment in the Balance: The Green Movement and the Supreme Court* (Cambridge, MA: Harvard University Press, 2015), pp. 289–290. On the matter of corruption in politics, Jane Mayer's *Dark Money* (New York: Doubleday, 2016) is superb, while Zephyr Teachout, *Corruption in America* (Cambridge, MA: Harvard University Press, 2014) provides historical depth and legal analysis. These two works should be read in tandom.

60 John Kenneth Galbraith, *The Economics of Innocent Fraud* (Boston: Houghton Mifflin Company, 2004), pp. 61–62. Psychologist Stephen Pinker argues otherwise in his cheerful assessment of human progress *The Better Angels of our Nature* (New York: Viking, 2011). Elizabeth Kolbert is not so sure, finding Pinker's case "exasperating" and selective—see her "Peace in Our Time," *New Yorker* (October 3, 2011). Pinker's thesis will be sorely tested in the hotter, more contentious, nuclear-armed years of the long emergency. James Madison, Speech in the Federal Convention (1787), in *James Madison: Writings*, ed. Jack Rakove (New York: The Library of America, 1999), p. 116; James Madison, "Political Observations," April 20, 1795, in *Letters and Other Writings of James Madison*, vol. 4, cited in David Unger, *The Emergency State* (New York: Penguin Books, 2012), p. 313.

61 James Madison, "Political Observations," April 20, 1795. See also Jill Lapore, "The Force," *New Yorker* (January 28, 2013): 70–76; Hugh Gusterson, "Empire of Bases," *Bulletin of the Atomic Scientists* (March 10, 2009); Michael Glennon, *National Security and Double Government* (New York: Oxford University Press, 2015), p. 7. Glennon's analysis is similar to that of Walter Bagehot's distinction between the "dignified" and "efficient" parts of the British government; one "intended to impress the many, the other efficient and intended to govern the many"—see Bagehot, *The English Constitution* (1867; Ithaca, NY: Cornell University Press, 1976), pp. 16, 22, 118, 176.

62 Andrew J. Bacevich, *The Limits of Power* (New York: Metropolitan Books, 2008), p. 101. This conclusion was broadly endorsed by political scientist Chalmers Johnson in his books *Nemesis* (New York: Metropolitan Books, 2006); *Sorrows of Empire* (New York: Metropolitan Books, 2004); and *Dismantling the Empire* (New York: Metropolitan Books, 2010).

63 Philip Taubman, *The Partnership* (New York: Harper, 2012), note 45.

64 Andrew Bard Schmookler, *Parable of the Tribes* (Berkeley: University of California Press, 1984) is a brilliant analysis of the underlying dynamics of military competition in a world of fragmented tribes or nation-states.

65 Jonathan Schell, *The Unconquerable World* (New York: Metropolitan Books, 2003), pp. 345, 357; see also Schell, "The Unfinished Twentieth Century," *Harper's Magazine* (January 2000): 41–56; Schell, *The Abolition* (1984; Stanford, CA: Stanford University Press, 2000); Schell, *The Seventh Decade* (New York: Metropolitan Books, 2007), pp. 216–217; and P. W. Singer, *Wired for War* (New York: Penguin Books, 2009). A world of smart drones, battlefield robots, and more are on the way as Annie Jacobsen describes in *The Pentagon's Brain* (New York: Little Brown, 2015).

66 The underside of U.S. foreign policy is replete with regrettable episodes with long-lasting consequences that seldom trouble the national conscience. The uncontained anti-communist zeal of the Dulles brothers in the 1950s is particularly tawdry as told in Stephen Kinzer, *The Brothers: John Foster Dulles, Allen Dulles, and Their Secret World War* (New York: Times Books, 2013); and David Talbot, *The Devil's Chessboard* (New York: Harper Collins, 2015).

67 Chris Hedges, *War Is a Force that Gives Us Meaning* (New York: Public Affairs, 2002), p. 3.

68 Edward O. Wilson, looking over our emergence as *Homo sapiens*, concludes that "we are addicted to tribal conflict." Wilson, *The Meaning of Human Existence* (New York: Liveright, 2014), p. 177. Quoted in Mark Kurlansky, *Nonviolence* (New York: Modern Library, 2006), p. 182.

69 Peter Ackerman and Jack Duvall, *A Force More Powerful* (New York: Palgrave, 2000). See also Anders Boserup and Andrew Mack, *War without Weapons* (New York: Schocken Books, 1976); and the lifetime work of Gene Sharp, including his *Waging Nonviolent Struggle* (Boston: Porter Sargent, 2005).

70 Edmund Russell, *War and Nature* (Cambridge, Eng.: Cambridge University Press, 2001) is the story of the transition from wartime chemicals for killing people to a profitable war on insects, weeds, and the small things that run the world.

71 Jeremy Rifkin, *The Zero Marginal Cost Society* (New York: Palgrave, 2014), pp. 11, 13.

72 George Zarkadakis, in his book *In Our Own Image* (New York: Pegasus Books, 2015), pp. 252–253, asks, "Have we stopped to consider the

vulnerability of such embedded computer systems to computer viruses, spies, terrorists, pranksters, and whomever might wish to access our data for nefarious goals?" and "What would the big data economy and the internet of things mean for politics and democracy?"

73 Zeynep Tufekci, "Smart Objects, Dumb Risks," *New York Times*, August 11, 2015; also Samuel Greengard, *The Internet of Things* (Cambridge, MA: MIT Press, 2015), pp. 134–165.

74 At the systems level, the work of the Next System Project is seminal. See Gar Alperovitz, James Gustave Speth, and Joe Guinan, *The Next System Project* (Takoma Park, MD: Next System Project, 2015); James Gustave Speth, *Getting to the Next System* (Takoma Park, MD: Next System Project, 2015). At the level of policy read Robert Reich, *Saving Capitalism* (New York: Knopf, 2015) and Joseph Stiglitz, *Rewriting the Rules of the American Economy* (New York: Norton, 2016).

Chapter 10. Sustainable Democracy?

Epigraphs. Edmund Burke, *Reflections on the Revolution in France* (1790; London: Penguin, 1986), pp. 194–195; Tony Judt, *Ill Fares the Land* (New York: Penguin, 2010), p. 3.

1 Lewis Mumford, *The Myth of the Machine: The Pentagon of Power* (New York: Harcourt Brace Jovanovich, 1970), pp. 433–435.

2 See Frank Kalinowski, *Environmental Legacies* (New York: Palgrave-Macmillan, 2016), chapters 7 and 8, in which he concludes that "the written Constitution of the United States is a document that has probably outlived its practical usefulness."

3 John Taylor Gatto, *A Different Kind of Teacher* (Berkeley, CA: Berkeley Hills Books, 2002).

4 Leopold Kohr, *The Breakdown of Nations* (1957; New York: Dutton, 1978), pp. 70, 79.

5 F. A. Hayek, *The Road to Serfdom* (1944; Chicago: University of Chicago Press, 2007), p. 234.

6 Frank Bryan, foreword to Susan Clark and Woden Teachout, *Slow Democracy* (White River Junction, VT: Chelsea Green, 2012), p. viii; John Dewey, *The Public and Its Problems* (1927; Chicago: Swallow Press, 1954), p. 213; Gar Alperovitz, *America beyond Capitalism* (Takoma Park, MD: Democracy Collaborative, 2011), p. 233; Robert Dahl and Edward Tufte, *Size and Democracy* (Stanford, CA: Stanford University Press, 1973), p. 140.

7 Clark and Teachout, *Slow Democracy*, p. 130.

8 Frank M. Bryan, *Real Democracy* (Chicago: University of Chicago Press, 2004), pp. 288–289.

9 Robert A. Dahl, *A Preface to Economic Democracy* (Berkeley: University of California Press, 1985), pp. 134–135.

10 Beth Simone Noveck, *Smart Citizens, Smarter State* (Cambridge, MA: Harvard University Press, 2015), pp. 78, 242–246, 262.

11 It was not always so. John Nichols traces Abraham Lincoln's concern for the rights of labor over those of capital to his reading of Karl Marx. See Nichols, *The "S" Word* (New York: Verso, 2011), pp. 61–99. Lincoln's Party has devolved.

12 Gar Alperovitz, *America beyond Capitalism* (Takoma Park, MD: Democracy Collaborative Press, 2011), p. 234; James Gustave Speth, *America the Possible* (New Haven: Yale University Press, 2012); Gar Alperovitz, James Gustave Speth, and Joe Guinan, *The Next System Project: New Political-Economic Possibilities for the 21st Century* (Takoma Park, MD: Next System Project, 2015).

13 Michael Shuman, *The Local Economy Solution* (White River Junction, VT: Chelsea Green, 2015) is excellent, as are his earlier books on building robust local economies. From the national perspective, see Robert Reich, *Saving Capitalism: For the Many, Not the Few* (New York: Knopf, 2015); and Joseph Stiglitz, *Rewriting the Rules of the American Economy* (New York: Norton, 2015).

14 Truth be told, I was once the proud owner of a Ford convertible equipped with a four-barrel carburetor with some 350 horsepower that was a considerable source of thrills, status, and pollution. In my more advanced age and with a more enlightened mind, I drive a ten-year-old Toyota Prius. Such is moral progress.

15 Sue Halpern, "Who Was Steve Jobs?," *New York Review of Books*, January 12, 2012.

16 Natasha Singer, "Can't Put Down Your Device? That's by Design," *New York Times*, December 5, 2015.

17 Natasha Dow Schüll, *Addiction by Design* (Princeton, NJ: Princeton University Press, 2012), pp. 77, 18, 29, 35.

18 Stuart Walker, *Designing Sustainability* (London: Routledge, 2014), pp. 35, 47, 45; also Victor Papenek, *Design for the Real World*, 2nd ed. (1984; Chicago: Academy Chicago Publishers, 1992), pp. 252, 293–299; Robert Grudin, *Design and Truth* (New Haven: Yale University Press, 2010), p. 131.

19 Steven H. Miles, *The Hippocratic Oath and the Ethics of Medicine* (New York: Oxford, 2004). As Miles explains, the Hippocratic Oath was more than a set of principles. In the world of around 400 BCE, it was regarded as more binding than a mere verbal promise and, in Miles's telling, something like a social institution. If violated, it involved perjury before the gods, which one may presume is not a good thing to commit (pp. 161–168). M. W. Thring, *The Engineer's Conscience* (London: Ipswich, 1992), p. 232; Seaton Baxter, "Deep Design and the Engineers Conscience," 2005, unpublished ms.

20 Architects Sim van der Ryn and John Tillman Lyle, along with biologists John and Nancy Todd, are pioneers in ecological design. See William McDonough and Michael Braungart, *The Upcycle* (New York: North Point Press, 2013), p. 10; Jay Harman, *The Shark's Paintbrush* (Ashland, OR: White Cloud Press, 2013); Grudin, *Design and Truth*, pp. 28–29.

21 Leonard Read, "I, Pencil" (Irvington-on-Hudson, NY: Foundation for Economic Education, 1958).

22 Randolph T. Hester, *Design for Ecological Democracy* (Cambridge, MA: MIT Press, 2006) is a thorough guide to "ecological democracy" and the use of design to rebuild the sinews of a coherent, participatory, and therefore resilient society.

23 Hillary Brown, *Next Generation Infrastructure* (Washington, DC: Island Press, 2014), pp. 38–39, also pp. 69–96 on "soft-path" water infrastructure.

24 Jared Diamond, *Collapse* (New York: Viking, 2005); Sue Roaf et al., *Adapting Buildings and Cities for Climate Change*, 2nd ed. (London: Elsevier, 2009); Alisdair McGregor et al., *Two Degrees: The Built Environment and Our Changing Climate* (London: Routledge, 2013); and Lewis Dartnell, *The Knowledge: How to Rebuild Civilization in the Aftermath of a Cataclysm* (New York: Penguin, 2014).

25 Roaf et al., *Adapting Buildings and Cities*, p. 344.

Chapter 11. Cities in a Hotter Time

Epigraphs. Kenneth Boulding, *The World as a Total System* (Beverly Hills, CA: Sage Press, 1985), p. 9; Donella Meadows, *Thinking in Systems* (White River Junction, VT: Chelsea Green, 2008), p. 11; Robert Jervis, *Systems Effects* (Princeton, NJ: Princeton University Press, 1997), p. 5; Garrett Hardin, "The Cybernetics of Competition," *Perspectives in Biology and Medicine* 7 (Autumn 1963): p. 77.

1 Jaime Lerner, *Urban Acupuncture* (Washington, DC: Island Press, 2014).

2 See Elizabeth Kolbert, "The Siege of Miami," *New Yorker* (December 21 and 28, 2015): 42–50, for a glimpse of the wetter reality.

3 The best compilation of writings on systems science is *Ecosystemology*, an unpublished reader assembled by Arnold Schultz for his classes at the University of California. See also Walter Buckley, ed., *Modern Systems Research for the Behavioral Scientist* (Chicago: Aldine, 1968).

4 Edward Hoagland, "What Would Aesop Think about What We're Doing to the Planet?" *New York Times*, March 24, 2013.

5 Richard Lazarus, "Super Wicked Problems and Climate Change," *Cornell Law Review* 94 (2009): 1153–1234, note 249; Lynton Keith Caldwell, *The National Environmental Policy Act* (Bloomington: Indiana University Press, 1998). Caldwell, the author of the act, was aware of its shortcomings and proposed that it be superseded by a Constitutional Amendment (pp. 147, 160–165).

6 Stephen Kellert and Judith Heerwagen, *Biophilic Design* (New York: John Wiley & Sons, 2008); and William Braham, *Architecture and Systems Ecology* (London: Routledge, 2016), which is an insightful union of the systems ecology of Howard and Eugene Odum with architecture.

7 Walter B. Cannon, *The Wisdom of the Body* (1932; New York: Norton, 1963). Before others, Cannon speculated that there were "general principles of stabilization" in industrial, domestic, social, and political affairs more or less isomorphic to those of the body. Sherwin Nuland, *The Wisdom of the Body* (New York: Knopf, 1977), pp. 355–356.

8 For example, Ted Kaptchuk, *The Web That Has No Weaver* (New York: Contemporary Books, 2000); and Dennis Normile, "The New Face of Traditional Chinese Medicine," *Science* 299 (January 10, 2003), pp. 188–190.

9 Jane Jacobs, *The Death and Life of Great American Cities* (New York: Vintage Books, 1961), p. 433.

10 See Bruce Katz and Jennifer Bradley, *The Metropolitan Revolution* (Washington, DC: Brookings Institution, 2013), pp. 1–13; Timon McPhearson, "Wicked Problems, Social-Ecological Systems, and the Utility of Systems Thinking," www.thenatureofcities.com, January 20, 2013; Peter Senge, *The Fifth Discipline* (1990; New York: Doubleday, 2006), p. 69; Meadows, *Thinking in Systems*, pp. 167–168.

11 Meadows, *Thinking in Systems*, p. 170.

12 Senge, *Fifth Discipline*, p. 78.

13 Carbon Neutral Cities Alliance, *Framework for Long-Term Deep Carbon Reduction Planning* (December 2015); Benjamin Barber, *Strong Democracy*

(Berkeley: University of Calilfornia Press, 1984), p. 227; James Fallows, "Why Cities Work Even When Washington Doesn't: The Case for Strong Mayors," *The Atlantic* (April 2014): 66–72; C40 Cities and ARUP, *Climate Action in Megacities 3.0* (December 2015). Estimates of reductions of CO_2 and the potential vary. Data gathered by the C40 staff estimates that reductions through 2015 are "10% relative to emission levels over a 2015 baseline," with the potential for considerably higher reductions by 2030 (p. 91). The Compact of Mayors is a network of 360 cities worldwide that aim to provide 25 percent of the gap between national climate targets and that necessary to hold the line at a warming of two degrees Celsius; the Carbon Neutral Cities Alliance of seventeen U.S. and global cities aims for reductions of 80 percent by 2050 "or sooner." See also Andy Gouldson et al., "Accelerating Low-Carbon Development in the World's Cities" (2015), available online at http://2015.newclimateeconomy.report/wp-content/uploads/2015/09/NCE2015_workingpaper_cities_final_web.pdf (accessed February 25, 2016).

14 Stephen Lansing and William Clark, *Priests and Programmers* (Princeton, NJ: Princeton University Press, 2007) is a brilliant cautionary tale about the interface of the presumptions of science and technology and old ways of allocating resources and managing complexity.

Chapter 12. The Oberlin Project

Epigraph. Petroleum V. Nasby was the nom de plume of humorist David Ross Locke. See Locke, *The Struggles (Social, Financial, and Political) of Petroleum V. Nasby* (Boston: I. N. Richardson, 1873) quoted in Nat Brandt, *The Town That Started the Civil War* (Syracuse, NY: Syracuse University Press, 1990), p. xiii.

1 John Kurtz, *John Frederick Oberlin* (Boulder, CO: Westview Press, 1976); J. Brent Morris, *Oberlin: Hotbed of Abolitionism* (Chapel Hill: University of North Carolina Press, 2014); Brandt, *The Town That Started the Civil War.*

2 Paul Crutzen, quoted in Vaclav Smil, *Harvesting the Biosphere* (Cambridge, MA: MIT Press, 2013), p. 237.

3 Donella Meadows, *Thinking in Systems* (White River Junction, VT: Chelsea Green, 2008).

4 One of my colleagues at Oberlin College, John Petersen, is a pioneer in the development and deployment of dashboards in public buildings,

schools, colleges, and businesses. He is a co-founder of Lucid Designs, Inc. See J. E. Petersen, C. Frantz, and M. R. Shammin, "Using Sociotechnical Feedback to Engage, Educate, Motivate and Empower Environmental Thought and Action," *Solutions* 5, no. 1 (2014): 79–87.

5 Peter Senge, *The Fifth Discipline* (New York: Doubleday, 2006).

6 David W. Orr, "Security by Design," *Solutions* (January–February 2012).

7 The Regional Planning Association included Lewis Mumford, Benton Mackaye, and others prominent in planning and land use issues; see Colin Woodard, *American Nations* (New York: Penguin, 2011).

INDEX

Ackerman, Bruce, 92–93, 223
Ackerman, Diane, 15
action networks, 89–90
Adams, Henry, 142
addiction, 20, 26, 69, 196–97
Addiction by Design (Schüll), 196
advertising, 20–21, 240n8,
    255n42
affection, 112–15
agriculture: Amish, 66–68;
    industrial, 23, 63, 209; and
    systems management, 208–9,
    210. *See also* food systems
Alaska Permanent Fund, 271
Alcoa Corporation, 219
Aldo Leopold Center, 163
Alperovitz, Gar, 189, 192–93
alternative energy, 14, 48–49, 101,
    144–45, 162
Altieri, Miguel, 208
American Association for
    Sustainability in Higher
    Education, 162
American dream, 29
American Scientist, 14
Amis, Martin, 139
Amish people, 66–69, 254n34
Anthropocene age, 5, 221
Apple, 195–96
Archer, David, 36
Arendt, Hannah, 104, 263n13
Arrhenius, Svante, 35

artificial intelligence (AI)
    systems, 8–11, 27, 79, 156–57,
    237n11, 238n16, 256n5, 272n32
Asimov, Isaac, 157
Atwood, Margaret, 118, 128
automobile industry, 194–95

Bacevich, Andrew, 177
Bacon, Francis, 23–24, 62, 63, 102,
    207, 238n16
Bailey, Liberty Hyde, 208
Bandes, Salo, 116–17
Barber, Benjamin, 92, 215
Barnes, Peter, 172
Barrat, James, 9
Barry, Brian, 76
Bateson, Gregory, 158, 273n38
Bauman, Zygmunt, 196–97
Baxter, Seaton, 198
Beatley, Tim, 150
Becker, Gary, 47–48
behavior, thoughts, and feelings,
    importance of changing, 19–20,
    33–34, 157–63
Berggruen, Nicolas, 32–33
Berkebile, Bob, 209
Bernays, Edward, 20, 21, 197
Berry, Thomas, 43, 86, 97,
    148–49, 158–59, 160, 258n25,
    270n16
Berry, Wendell, 101, 208–9,
    247n10, 252n22